Microbiology Concepts with Experiments for Agricultural and Food Microbiology

Microbiology Concepts with Experiments for Agricultural and Food Microbiology

Khushboo Chaudhary and
Pankaj Kumar Saraswat

Delve
PUBLISHING
www.delvepublishing.com

Microbiology Concepts with Experiments for Agricultural and Food Microbiology

Khushboo Chaudhary and Pankaj Kumar Saraswat

Delve Publishing
224 Shoreacres Road
Burlington, ON L7L 2H2
Canada
www.delvepublishing.com
Email: orders@arclereducation.com

ISBN: 978-1-77469-188-5 (Hardcover)

ABOUT THE AUTHORS

Dr Khushboo Chaudhary is presently working as a Research Associate in NRCE, Hisar Haryana, India and having one year of teaching experience and seven years of research experience. Previously, she worked on "Improvement of Phytoremediation efficiency of Fluoride". She has published several research papers in international and national journals. She has published five international text books. She has got seven best paper and poster presentation awardsfromthe Indian Society of Genetics and Biotechnology Research and Development and received the president appreciation awards also in the International conference. She has got the best poster award by ISSGPU Central Institute Research on Goats, Makhdoom. She has published several gene banks in NCBI Pubmed. She has also published a research article in the virology journal. She is likely to be a co-author in several of the publications and coauthor in J. Virological Methods.

Dr.Pankaj Kumar Saraswat graduated from R.B.S. College Bichpuri (Agra University) Agra in 1996, completed post-graduate and Ph.D. in Soil Science & Agricultural Chemistry from Banaras Hindu University Varanasi in 1999 and 2004 respectively. Dr. Saraswat started his career at H.N.B.G. University Srinagar Garhwal Uttrakhand in 2005 as a lecturer in Soil Science and moved as Subject Matter Specialist (Soil Science) to KVK Banasthali Vidyapith Tonk Rajasthan. In October 2015 Dr. Saraswat joined ICAR-RC for N.E.H Region Umiam as Sr. Scientist & Head at KVK Tamenglong Manipur. Dr. Saraswat as P.I. completed two research projects externally funded from SERC-DST Govt. of India New Delhi and DST-Govt. of Rajasthan Jaipur. At the same time, he also conducted four INSPIRE-Internship Science Camps under the SEAT program of DST for 10th pass top 1% students of Rajasthan. He has published three international text books Presently Dr. Saraswat has been working for agricultural technology assessment and demonstrations for its wider application at farmers' field along with capacity development programs for farmers, farm women, rural youth, line departments and other stakeholders and also working as P.I. in one of NEC Shillong funded demonstration based project in Tamenglong district Imphal, Manipur.

TABLE OF CONTENTS

PREFACE

Microbiology Concepts with Experiments for Agricultural and Food Microbiology has always been a fundamental area of science. The exponential rise in the quantity of scientific information and the rate at which new discoveries are made, require much elaborated, interdisciplinary and up-to date information and their understanding. The book "Microbiology Concepts with Experiments for Agricultural and Food Microbiology" has been written to serve as a guidebook for undergraduates and postgraduates of different universities. This guidebook has covered a range of syllabus and the student has an easy way to go through this book content. Books are available on the subject but still, students face problems to find notes. While this book is for students to acquire or gain fair knowledge and easy presentation of language has written for proper understanding of each and every topic. The chapters have been so arranged to give a sequential knowledge of the introductory basic agricultural microbiology, molecular biology, plant metabolic pathways, elements of immunology, industrial microbiology, fermentation technology and principle working of basic equipments in this guide book. For preparing this book, I have been greatly helped by several books and presentations which I express my acknowledgment in the references and if we have forgotten any references in my notice, please ignore them. The numbers of scientists and authors have been quoted or fully acknowledged in the bibliography at the end of the book. We claim no originality for the matter or text presented in this book but the method of presentation and illustration is our own. Last but not least I would be thankful to readers for pointing out errors and most welcome suggestions in the next guide book.

Khushboo Chaudhary

Pankaj Kumar Saraswat

Chapter 1

AGRICULTURAL MICROBIOLOGY

1. HISTORY OF MICROBIOLOGY: SPONTANEOUS GENERATION THEORY

The study of microbes or microorganisms is called microbiology. The microbes, or microorganisms are minute living things that are usually unable to be viewed with the naked eye. Some examples of microbes are as follows: Bacteria, fungi, protozoa, algae, viruses, and some are parasites (helminths). Some are pathogenic and some are not. Most bacterial species cannot cause disease

Beneficial Role of Microbes

- Production of antibiotics and foodstuffs.
- Decompose organic waste.
- Produce industrial chemicals such as ethyl alcohol and acetone.
- Produce fermented foods such as vinegar, cheese, and bread.

Ancient Microbiology

Ancient people recognized many factors involved in diseases. Most ancient people documented that some diseases are communicable, for example: When black death struck Europe entire villages were abandoned in an effort to escape the highly infectious plague (1347 A.D). No medical knowledge existed in Medieval England to cope with the disease. The Romans were a hygienic bunch and were concerned with health and cleanliness (waste and sewage). The Romans invented the first underground sewage system. The Romans understood that sewage could cause disease , therefore, decided to build an underground sewage system which is an idea we still use today. A network of pipes brought clean water into the city of Rome and removed waste. Waste flushed from the latrines flowed through a central channel into the main sewage system and thence into a nearby river or stream. The public bath houses were the place where people went to socialize and do business as well as getting clean.

History of Microbiology

First microbes were observed in 1673. **Robert Hooke-** in 1665 reported that living things were composed of cells. The devise of the compound microscope and illumination system are his.

Antoni van Leeuwenhoek- (1673-1723): He is considered the father of bacteriology. He made simple microscopes and began observing with them (1674). He discovered bacteria (he called them animalcules). Leeuwenhoek's microscope:

- A screw for adjusting the height of the object being examined
- A metal plate serving as the body
- A skewer to impale the object and rotate it
- The lens itself, which was spherical

Spontaneous Generation

Aristotle synthesized the hypothesis which stated that some vital force contained in given organic matter can create living organisms from inanimate objects. In basic terms, spontaneous generation stated that living organisms arise from nonliving matter. Spontaneous generation was disproved in 1668 by Italian Scientist, **Francesco Redi**.

Redi's Experiment: Where do maggots come from?

Hypothesized: Redi suspected that flies landing on the meat laid eggs that eventually grew into maggots

Experiment: Placed meat in three separate jars

Disproving Spontaneous Generation: Redi's Experiment

Jar #1	Jar #2	Jar #3
Left open Maggots developed, Flies were observed laying eggs on the meat in the open jar	**Covered with netting** Maggots appeared on the netting and Flies were observed laying eggs on the netting	**Sealed** No maggots developed

Spontaneous Generation

John Needham- revived the theory of spontaneous generation in 1745. **Lazzaro Spallanzani (1729-1799)**. In 1776 he demonstrated that microorganisms were already in the solution, the container, or the air. He took solutions that he knew would "breed" organisms and **boiled** them for up to an hour. The flasks were hermetically sealed to keep out contaminated air.

Biogenesis: In 1858 German scientist, **Rudolf Virchow** challenged spontaneous generation with his concept of **biogenesis.** Living organisms arise from pre-existing life. Virchow presented his idea to the scientific community, but could not back it up with a convincing experiment.

In 1861, a French scientist by the name of **Louis Pasteur** demonstrated where microorganisms came from-he is the Father of medical microbiology: he demonstrated that microorganisms exist in the air and could contaminate sterile solutions by passing air through cotton filters. The filter trapped tiny particles floating in the air.**Louis Pasteur**

Louis Pasteur performed numerous experiments to discover why wine and dairy products became sour. He found that bacteria were to blame (lactic acid fermentation). Pasteur called attention to the importance of microorganisms in everyday life and stirred scientists to think that if bacteria could make the wine "sick," then perhaps they could cause human illness.

The Golden Age of Microbiology 1857- 1914

Beginning with Pasteur's work, discoveries included the relationship between microbes and disease, immunity and antimicrobial medicine

Germ Theory of Disease

Germ theory states that specific microscopic organisms are the cause of specific diseases.

Pasteur And Germ Theory of Disease

Pasteur showed that microbes are responsible for fermentation. Microbial growth is also responsible for the spoilage of food. Pasteur demonstrated that spoilage bacteria could be killed with heat (**pasteurization**). It was discovered that weak forms of disease could be used as an immunization against stronger forms and that rabies was transmitted by viruses too small to be seen under the microscopes of the time. He developed vaccines for anthrax (1881) and rabies (1885).

History Of Vaccination

A precursor of smallpox **vaccination** was variolation. The Germ theory of disease was not known at this time. An early Asian method introduced dried scabs of smallpox patients and was later modified in Europe. The modification consisted of injecting infectious material under the skin. First tested among abandoned children and prisoners when it was declared safe, members of the English royal family were inoculated.

Smallpox Vaccine

In 1796: Edward Jenner is credited with the development of the smallpox vaccine. Folk wisdom suggested that dairy maids who had contracted cowpox seemed to be immune to smallpox. Infection with the cowpox virus produced a much less severe form of the disease than smallpox. Jenner conducted an experiment in which he used scabs from the cowpox lesions on the arm of a dairy maid, Sarah Nelmes to create a small pox vaccine. He then used the material to vaccinate an 8-year-old boy, James Phipps. After being vaccinated Phipps appeared to develop immunity to the smallpox virus. Later Jenner vaccinated his own son and several other children. He obtained similar results. Worldwide elimination of smallpox was achieved in 1978. He called vaccination from vacca for cow. The protection is called immunity

Discovery of Penicillin

In 1928 Alexander Fleming discovered the first antibiotic. He observed that Penicillium fungus made an antibiotic, penicillin, that killed S. aureus. In the 1940s Penicillin was tested clinically and mass produced and was available towards the end of World War II.

Development of Agar

Angelina Hesse developed the use of agar to grow microorganisms. She was the wife of Walter Hesse who worked in Koch's laboratory. The benefit of agar was it was not attacked by most bacteria. Agar is better than gelatin because of its higher melting point (96°c) and solidifying (40–45°c) points.

2. ROLE OF MICROBES IN FERMENTATION

Introduction

The fermentation industry is composed of five major bio-ingredient categories. They are proteins & amino acids, organic acids, antibiotics, enzymes and vitamins & hormones.

Fermentation Industry

- The cost and availability of feed-stocks are derived.
- The efficiency of industrial microorganisms.
- Fermentation condition and optimization.
- Down stream process and end-product recovery efficiency.
- Fermentation by-product utilization.
- Utility consumption and labor cost.

Fermentation Media

Optimum balance of the media is mandatory for cells propagation and for the maximum production of target metabolites.

Media Compositions

- Carbon source
- Nitrogen source
- Minerals
- Growth factors
- Mutants

Industrial Microorganisms

- Microbial screening (Wild strains).
- Microbial yield improvement (Mutation).
- Recombinant DNA (Genetically engineered).
- Microbial selection.
- Industrial microorganism.

Types of Fermentation

- Solid state fermentation
- Liquid state fermentation surface culture & submerged culture

Solid State Fermentation

- Solid state fermentation process can be defined as microbial growth on particles without the presence of free water.
- Particles are a solid culture substrate such as rice or wheat bran saturated with water and inoculated with (mold, yeast, bacteria) at controlled room temperature.
- It is ideal for growing filamentous fungi.
- It has been used in Asia and developing nations.
- It is more cost-effective (smaller vessels lower water consumption, reduced waste water treatment costs, lower energy consumption, and lesser contamination problems).

Applications

Potentially many high-value products such as extra-cellular enzymes, primary metabolites, and antibiotics could be produced in solid-state fermentation. It is estimated that nearly a third of industrial enzyme produced in Japan is made by SSF process. Production of organic and ethanol from starch substrates: the digestibility of fibers and lignocelluloses materials for both human and animal consumption.

Liquid State Fermentation: Submerged Culture

The submerged culture is performed in tanks that can reach in size for over 100,000 gallons. It is ideal for growing unicellular organisms such as bacteria and yeast.

Liquid state fermentation methods

- Batch fermentation.
- Fed-batch fermentation.
- Continuous fermentation.
- Semi-continuous fermentation.

Batch fermentation

It is considered to be a closed system. The sterilized media in the fermenter is inoculated with the microorganism. Incubation is allowed under the optimum conditions (aeration, agitation and temperature). The during entire fermentation nothing is added except air, antifoam and acid/base.

Fed-batch fermentation

It is an enhancement of batch fermentation with continous adding the nutrients (feeding) in a small dose during the fermentation. The method in controlling the nutrients feeding process is by measuring methods. The main advantage of fed-batch fermentation is the elimination of catabolite repression (feed-back inhibition).

Continuous fermentation

It is an open system and continuously sterile nutrient is added and the converted nutrient is taken out from the fermentor. In continuous process cell loss as a result of outflow must be balanced by the growth of the microorganism.

Important factors for continuous fermentation

The system must be stable for at least 500 hours and maintaining sterile conditions for all periods of fermentation time. The composition of nutrients must be constant all the time. It is maintaining the strain stability for constant high production yield.

Semi-continuous fermentation

Semi-continuous fermentations, in which a fraction of the fermentation medium is replaced with fresh media at regular intervals.

Enzymes

Enzymes are active biological molecules responsible for thousands of metabolic processes that sustain life. Most enzymes are proteins, although some catalytic enzymes are RNA molecules. In enzyme reactions, the molecules at the beginning of the process called **substrates** are converted into different molecules called **end-products.** Enzymes are very specific as to which reaction they catalyze and the substrate involved in the reaction. It depends on enzyme activity, the bioconversion to end-products can be faster and reached the equilibrium state rapidly.

Factors of Enzyme

- **Temperature:** The optimum is generally 40-60^{0}C. Some enzymes exhibit an optimum at almost 100^{0}C.
- **Value of pH:** The optimum is generally in the range of 5-7. The extreme values of 1.5-10.5 have been found.
- **Activation:** Many chemicals activate the catalytic enzyme activity, such as inorganic ions.
- **Inhibitors:** Many chemicals inhibit the catalytic enzyme activity.
- There are two types of enzymes inhibition: Irreversible inhibitors (competitive inhibition) and reversible inhibitors (uncompetitive inhibition).
- **Substrate inhibition:** High concentration of substrate may inhibit the catalytic activity of an enzyme.
- **End-product inhibition**: In the case of multi-enzyme system (catalytic inhibition).

Activators Cofactors and Coenzymes

- Some enzymes do not need any additional components to show full activity.
- Cofactors can be either inorganic (metals) or organic compounds (flavin and heme).
- Coenzymes include NAD^+, $NADP^+$ and ATP.
- These coenzymes transfer chemical groups between enzymes.
- The chemical groups are carried by the hydride ion (H^+) by NADH or NADPH.

$NAD^+ + 2H^+ + 2e^-$ NADH / $NADP^+ + 2O^-_2 + 2H^+$ NADPH + $2O_2$

Or phosphate groups carried by ATP.

ATP +H_2o ADP+P_1 (-7.3kcal/ mole) / ATP +H_2o AMP + PP_1 (-14kcal/ mole)

Classification of Enzymes

The International of Biochemistry and Molecular Biology developed a nomenclature for enzymes (EC number). Each enzyme is classified by the sequence of four numbers led by EC. (E.C. 5.3.1.18 Glucose isomerase)

Classification

- **EC1** *Oxidoreductases* (catalyze oxidation/reduction reactions).
- **EC2** *Tranferases* (transfer a functional group).
- **EC3** *Hydrolases* (catalyze the hydrolysis of various bonds).
- **EC4** *Lyases* (cleave bonds by mean of hydrolysis /oxidation).
- **EC5** *Isomerases* (isomerization within same molecule).
- **EC6** *Ligase* (join two molecules with covalent bonds).

Enzymes Production

Constitutive enzymes: The microorganism produces the enzyme in minimal fermentation media.

Inducible enzymes: The microorganism requires adding inducible agents in the media to produce the target enzyme.

Extracellular enzymes: The microorganism secretes the enzyme in the fermentation media.

Intracellular enzymes: The microorganism produces the enzyme inside the cell.

Enzymes are usually sold based on the activity (u/ml or u/gm). If the efficiency of enzymes is considered, their cost, is based on active enzyme protein u/mg protein (specific activity). The commercial exploitation of enzymes ranges from high-volume but the low cost (industrial enzymes) to low volume, but high cost (enzymes for medical, scientific and analytical use). Workers handling industrial enzymes should use protective clothing and eye protection. Food enzymes in foods processing have been shown to be safe through many years of manufacturing practices.

3. GERM THEORY OF DISEASE

Introduction

Robert Koch in 1867 provided proof that bacterium causes anthrax and provided the experimental steps, Koch's postulates, used to prove that a specific microbe causes a specific disease. The pathogen must be present in all cases of diseases. The pathogen must be isolated and grown in a lab in pure culture. Pathogen from pure cultures must cause disease when inoculated into healthy, susceptible lab animals. The same pathogen must be isolated from the diseased lab animal. Trillion microbes (Bacteria, fungi and actinomycetes) fig. 3.1 are in/on the human body such as indigenous microflora, opportunistic pathogens, producers of vitamins, enzymes, infectious disease (caused by microbes) and microbial intoxication (caused by toxins of microbes).

Agostino Bassi in 1835 showed that silkworm diseases were caused by a fungus.

Pasteur in 1865 believed that another silkworm disease was caused by a protozoan.

Ignaz Semmelweis in 1840 advocated hand washing to prevent transmission of fever from one OB patient to another.

Joseph Lister in 1860, he the father of antiseptic surgery. He used a chemical disinfectant (carbolic acid) to prevent surgical wound infections after reading Pasteur's work showing microbes are in the air, can spoil food, and cause animal disease.

Microbes in the Environment

- Saprophytes: decompose materials
- Genetically engineered bioremediators
- Microbial ecology
- Food chains plankton

Phytoplankton

Zooplankton

- Biotech in food/beverages
- Antibiotics
- Photosynthesis: production of oxygen

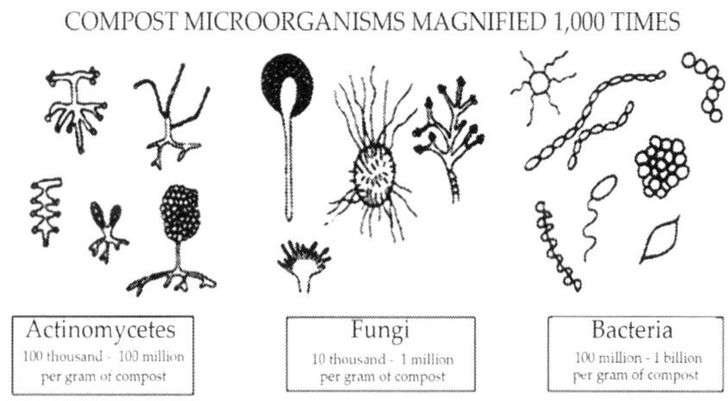

Figure 3.1. Microorganisms (Bacteria, fungi and actinomycetes).

Only a few hundred years old as a science. *Mycobacterium tuberculosis* (the causative agent of tuberculosis) has been found in the DNA of 3000-year-old mummies.

Anton van Leeuwenhoek (1632-1723): Animalcules (1678); live microbes and 3 basic shapes of bacteria

- Cocci
- Bacilli
- Spirilla

Anton van Leeuwenhoek Father of Microbiology, bacteriology, protozoology, microscopy and no disease connection suspected

Robert Hooke (1635-1703)

He observed slices of cork in 1665, England defined the "Cells" no stains. The beginning of the "cell theory" and all living things are composed of cells.

Spontaneous Generation

The theory held by most until the late 1800's, Abiogenesis, disproved and lead to theory, two scientists paved the way.

- Francesco Redi
- Lazzaro Spallanzani

Redi's experiment in 1668 meat in an open jar = flies and meat in a sealed jar or one with a cloth over it = no flies.

Spallanzani's Experiment – 1768

Sterilize solutions using heat. Seal one, leave the other open. The growth/ turbidity in open flask only.

Edward Jenner (1749-1823)

- Smallpox outbreaks
- Cowpox
- Cowpox vaccine
- Inoculation

Louis Pasteur (1822-1895)

- Pasteur institute
- French biochemist
- Ultimately proved biogenesis
- Germ theory of fermentation
- Germ theory of disease
- Aerobes and anaerobes
- Pasteurization
- Father of bacteriology & immunology
- Vaccines, attenuation and immunity

Koch's Postulates – 1884

It was used anthrax in sheep to develop, pathogen present in each case, It must be isolated from a host, grow in pure culture, reproduce disease from the culture in others, repeat about Tuberculosis, cholera, typhoid.

Exceptions to Koch's Postulates

It is limited to pathogens that may be cultured *in vitro*, therefore excludes viruses and certain unusual bacteria, obligate intracellular pathogens and fastidious microorganisms. Some pathogens change when grown *in vitro* instead of *in vivo*. **Antibiotics** were discovered Alexander Fleming, 1928.

Polio Vaccines

Injection (Salk, 1955)

Oral (Sabin, 1957)

Chicken Pox Vaccine

It was developed in 1974 and available in 1995 such as in *Haemophilus influenza* vaccine (HiB): Meningitis and Hepatitis B vaccinations. Luc Montagnier (1982) was discovered HIV and Robert Gallo (1984) HIV causes AIDS.

4. APPLIED AREAS OF MICROBIOLOGY

Introduction

Disease is a disturbance in the state of health. Microbes cause disease in the course of stealing space, nutrients, and/or living tissue from their symbiotic hosts. Microbes do most of the following: gain access to the host (contamination). Adhere to the host (adherence), replicate on the host (colonization), invade tissues (invasion), produce toxins or other agents that cause host harm (damage).

Characteristics

Infectivity – the ability to infect a host

Pathogenicity – the ability to cause disease in the host

Virulence – the ability to cause severe disease in the host

Immunogenicity –the ability to induce an immune response in the host

Infectious disease

Infectious dose is the number of organisms are needed to successfully infect the latent period is exposure to infectiousness interval. The incubation period is an interval from exposure to clinical symptoms. The infectious period is an interval during which the host can transmit infection. Reproductive rate is the ability of an agent to spread in populations. The outbreak is limited spread. Endemic is usually present; steady prevalence. An epidemic is rapid spread. Pandemic is occurring across countries and in multiple populations.

Modes of Transmission

Direct transmission through droplet, aerosol and skin to skin.

Indirect transmission through fomites (clothes, blankets, door handles etc), vectors (e.g. mosquitoes), food and water intermediate hosts (e.g. snails).

Clinical Responses

Inapparent infection has no clinical symptoms generated. The carrier state has usually no clinical symptoms but the host can transmit infection for long periods. The clinical symptoms are mild disease, severe disease, residual impairment and death.

- Bacteria are surviving on appropriate media, stain gram-positive or negative
- Viruses are obligate intracellular parasites that only replicate intracellularly (DNA and RNA).
- Fungi are non-motile filamentous and branching strands of connected cells.
- Metazoa is multicellular animals (e.g. parasites) with complicated life cycles often involving several hosts.
- Protozoa are single-cell organisms with a well-defined nucleus.
- Rickettsia are very small bacteria spread by ticks.
- Prions are unique proteins lacking genetic molecules.
- Chlamydia is bacteria lacking cell walls.

Bacteria are classified by their gram stain characteristics. Gram staining is the application of a crystal violet dye to a culture of bacteria. Bacteria that retain the color of the dye are called gram-positive; bacteria that don't are gram-negative fig. 4.1. The gram stain attaches to peptidoglycan in the bacterial cell wall. In gram-negative bacteria, the peptidoglycan layer is protected by an outer membrane.

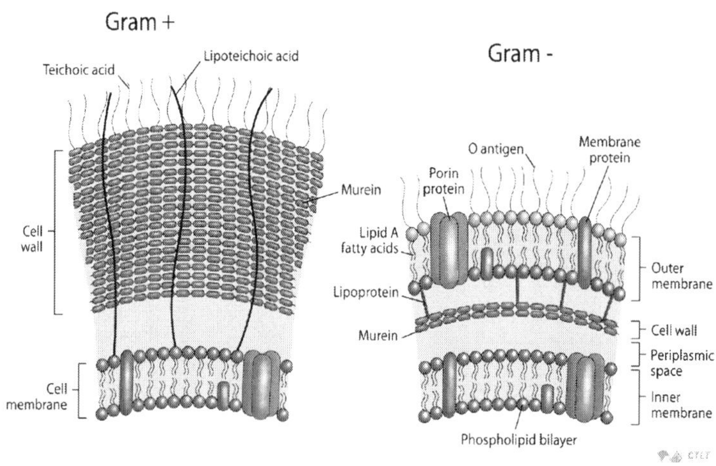

Figure 4.1. Gram positive and Gram negative bacteria.

Viruses are acellular, obligate intracellular organisms. The complete infectious virus is termed a virion. The virion consists of the specific nucleic acid (DNA or RNA) surrounded by a protein coat (capsid). Some viruses are enveloped which means that they possess a lipoprotein coat that surrounds the capsid and is acquired from the infected host cell membrane. Viruses that lack an envelope are "naked." Viruses are typically classified by: genetic material (DNA vs. RNA), strandedness (single vs. double) and size and shape of the capsid and whether it is enveloped or non-enveloped.

Method of Replication

All fungi are chemoheterotrophs. Pathogenic fungi have two forms: yeasts (unicellular) and molds (multicellular). Some fungi are dimorphic (this is particularly true for the pathogenic fungi). Molds grow as filamentous, branching strands of connected cells known as hyphae.

Types of Parasites

Protozoa: single-celled, microscopic organisms that can perform all necessary functions of metabolism and reproduction. Some protozoa are free-living, while others parasitize other organisms for their nutrients and life cycle. The morphology of protozoa varies widely and includes oval, spherical and elongated cells that can range in size from 5-10 to 1-2 mm. Structurally, the protozoa resemble other eukaryotic cells and possess a cytoplasmic membrane that encloses cytoplasm containing membrane-bound nuclei, mitochondria, 80s ribosomes and a variety of organelles. Helminths: A large, multicellular organism (worm) that is generally visible to the naked eye in its adult stages. Helminths can be free-living or parasitic Nematodes: Roundworms, Trematodes: Flukes and Cestodes: Tapeworms. Prions are abnormal, transmissible agents that are able to induce abnormal folding of normal cellular prion proteins in the brain, leading to brain damage and the characteristics signs and symptoms of the disease. Prion diseases are usually rapidly progressive and always fatal.

5. METABOLISM IN BACTERIA: ATP GENERATION, CHEMOAUTOTROPHY, PHOTO-AUTOTROPHY, PHOTO-RESPIRATION

Introduction

Bacterial photosynthesis is a light-dependent, anaerobic mode of metabolism. Carbon dioxide is reduced to glucose, which is used for both biosynthesis and energy production. Depending on the hydrogen source used to reduce CO_2, both photolithotrophic and photoorganotrophic reactions exist in bacteria. Bacteria require organic molecules to provide their carbon and energy. The energy-yielding catabolic reactions can be of many different types, although they all involve electron-transfer reactions in which the movement of an electron from one molecule to another is coupled with an energy-trapping reaction that yields ATP fig.5.1 and Table 5.1 .

- Phototroph uses light for energy.
- Chemotroph uses chemical reactions for energy.
- Autotroph takes in CO_2 for a carbon source.
- Heterotroph takes in organic compounds for a carbon source.
- Photoautotroph takes energy from light / carbon from atmosphere.
- Photoheterotroph is Energy from light / carbon from organic compounds.
- Chemoautotroph takes energy from chemical reactions (inorganic) / carbon from the atmosphere.
- Chemoheterotroph takes energy from organic compounds / carbon from organic compounds. It obtains energy from compounds - chemicals.
- Chemoheterotroph uses energy is used in anabolism. It obtains energy from chemicals. Carbon from the atmosphere not from compounds "eaten". Chemoautotroph, *Thiobacillus ferroxidans* and energy used in the Calvin-Benson cycle to fix CO_2.
- Phototrophs use light energy.
- Photoautotrophs can use the light energy in the Calvin-Benson cycle to fix CO_2 and make sugars to use for energy. Photoheterotrophs use light energy and organic compounds.

Table 5.1. Metabolism

Nutritional type	**Energy source**	**Carbon source**	**Example**
Photoautotroph	Light	CO_2	**Oxygenic**: Cyanobacteria plants. **Anoxygenic**: Green, purple bacteria.
Photoheterotroph	Light	Organic compounds	Green, purple nonsulfur bacteria.
Chemoautotroph	Chemical	CO_2	Iron-oxidizing bacteria.
Chemoheterotroph	Chemical	Organic compounds	Fermentative bacteria. Animals, protozoa, fungi, bacteria.

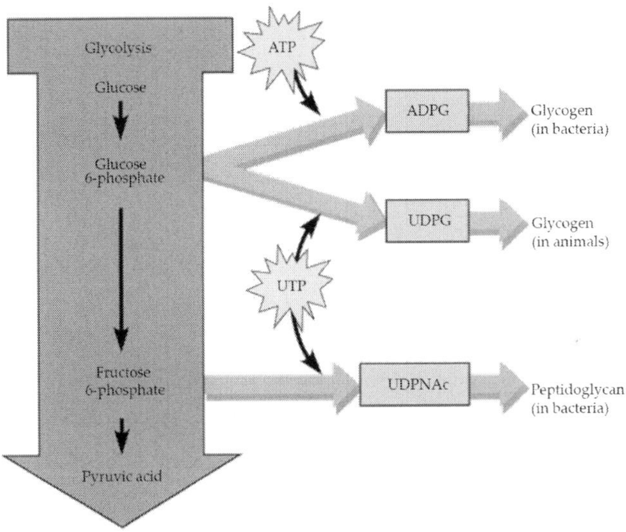

Figure 5.1. Metabolic pathways of energy use for polysaccharide biosynthesis .

Lipid metabolic pathways are lipid biosynthesis, lipid catabolism and beta-oxidation which uses 2 carbons at a time to make acetyl groups fig. 5.2 and 5.3.

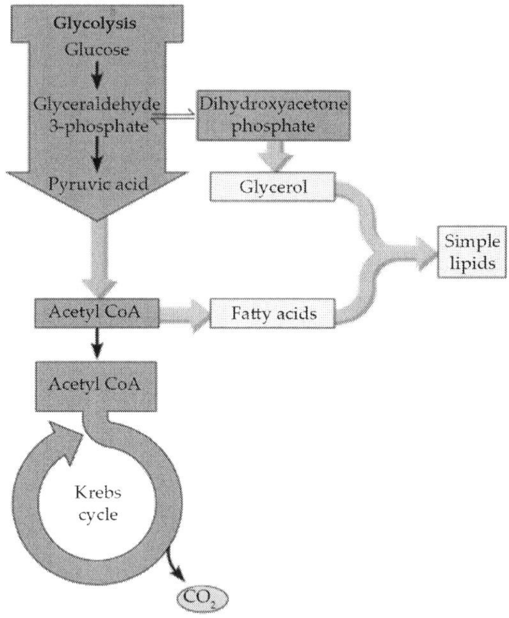

Figure 5.2. Amino acid and protein biosynthesis.

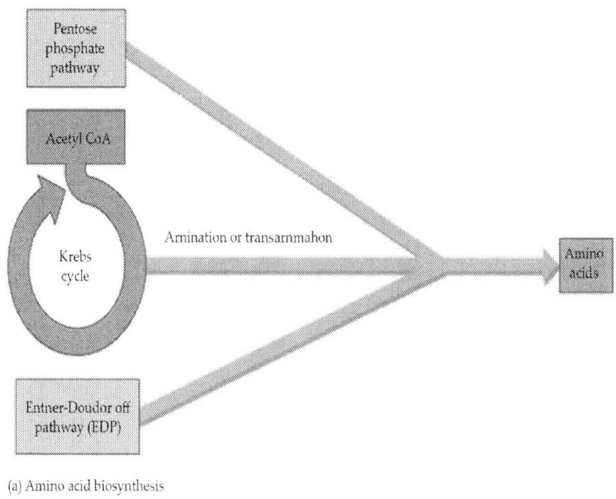

Figure 5.3. Amino acid biosynthesis.

6. LYSOGENIC CYCLES: VIROIDS AND PRIONS

Introduction

Viruses are both living or non-living. They have some properties of life but not others, for example, viruses can be killed, even crystallized like table salt. However, they can't maintain a constant internal state (homeostasis). A virus is a noncellular particle made up of genetic material (DNA) and protein that can invade living cells.

Discovery of Viruses

Beijerinck (1897) was coined the Latin name "virus" meaning poison. He studied filtered plant juices & found they caused healthy plants to become sick.

Tobacco Mosaic Virus: Wendell Stanley in 1935 was crystallized sap from sick tobacco plants and he discovered viruses were made of nucleic acid and protein

Smallpox: Edward Jenner in 1796 was developed a smallpox vaccine using milder cowpox viruses, Deadly viruses are said to be virulent, smallpox had been eradicated in the world today, but is now coming back. Viruses are smaller than the smallest cell, measured in nanometers, Viruses couldn't be seen until the electron microscope was invented in the 20th century.

Characteristics

Non living structures, noncellular, have a protein coat called the capsid. It has a nucleic acid core containing DNA or RNA. It is capable of reproducing only when inside a host cell. Some viruses are enclosed in a protective envelope. Some viruses may have spikes to help attach to the host cell. Most viruses infect only specific host cells. The capsids (coats) are made of individual protein subunits called capsomeres. Viruses are inactive outside of host cells, lack ribosomes and enzymes needed for metabolism. Use the raw materials and enzymes of the host cell to be able to reproduce. Some viruses cause diseases such as: smallpox, measles, mononucleosis, influenza, colds, warts, AIDS, Ebola. Some viruses may cause some cancers like leukemia. Virus-free cells are rare. Viruses come in a variety of shapes. Some may be in helical shape like the Ebola virus. Some may be polyhedral shapes like the influenza virus. Others have more complex shapes like bacteriophages are RNA or DNA virus do or do not have an envelope, capsid shape, host they infect. Viruses that attack bacteria are bacteriophages (phage): T-phages are a specific class of bacteriophages with icosahedral heads, double-stranded DNA, and tails. **T-phages** The most commonly studied T-phages are T4 and T7. They infect *E. coli*, intestinal bacteria. Six small spikes at the base of a contractile tail are used to attach to the host cell. Inject viral DNA into the cell. T-4 Bacteriophage head with 20 triangular surfaces, capsid contains DNA, head & tail fibers made of protein.

Retroviruses

It is containing RNA, not DNA. Family Retroviridae. Contain enzymes called Reverse Transcriptase. When a retrovirus infects a cell, it injects its RNA and reverse transcriptase enzyme into the cytoplasm of that cell. The enzyme reverse transcriptase (or RTase), which causes the synthesis of a complementary DNA molecule (cDNA) using virus RNA as a template. HIV or AIDS virus is a retrovirus. Feline Leukemia virus is also a retrovirus.

Viral Replication

Once a viral genome has entered a cell, the cell begins to manufacture viral proteins. The virus makes use of host enzymes, ribosomes, tRNAs, amino acids, ATP, and other molecules. Viral nucleic acid molecules and capsomeres spontaneously self-assemble into new viruses. Phages are the best understood of all viruses. Phages have two reproductive mechanisms: the lytic cycle and the lysogenic cycle.

Viral Attack

Is host specific. Humans rarely share viral diseases with other animals. Eukaryotic viruses usually have protective envelopes made from the host cell membrane.

Lytic Cycle

The lytic cycle is a phage replicative cycle that culminates in the death of the host cell. The lytic cycle produces new phages and lyses (breaks open) the host's cell wall, releasing the progeny viruses. A phage that reproduces only by the lytic cycle is called a virulent phage. Bacteria have defenses against phages, including restriction enzymes that recognize and cut up certain phage DNA.

Steps of the Lytic Cycle

Attachment to the cell, penetration (injection) of viral DNA or RNA, replication (Biosynthesis) of new viral proteins and nucleic acids, assembly (Maturation) of the new viruses, the release of the new viruses into the environment (cell lyses).

Lysogenic Cycle

The lysogenic cycle replicates the phage genome without destroying the host. The viral DNA molecule is incorporated into the host cell's chromosome. This integrated viral DNA is known as a prophage. Every time the host divides, it copies the phage DNA and passes the copies to daughter cells. An environmental signal can trigger the virus genome to exit the bacterial chromosome and switch to the lytic mode. Phages that use both the lytic and lysogenic cycles are called temperate phages.

Viral Latency

Some viruses have the ability to become dormant inside the cell called latent viruses. They may remain inactive for long periods of time (years). Later, they activate to produce new viruses in response to some external signal. HIV and Herpes viruses are examples. Once a prophage cell is activated, the host cell enters the lytic cell. New viruses form & the cell lyses (bursts).

Latency in Eukaryotes

Some eukaryotic viruses remain dormant for many years in the nervous system tissues. Chickenpox (caused by the virus *Varicella zoster*) is a childhood infection. It can reappear later in life as shingles, a painful itching rash limited to small areas of the body. Herpes viruses also become latent in the nervous system. A herpes infection lasts for a person's lifetime such as Genital herpes (Herpes Simplex 2), Cold sores or fever blisters (Herpes Simplex1).

Treatment for Viral Disease

Vaccines

An attenuated virus is a weakened, less vigorous virus. An attenuated virus is capable of stimulating an immune response and creating immunity, but not causing illness.

Viral Treatments

Interferons are naturally occurring proteins made by cells to fight viruses. Antiviral drugs (AZT), Protease inhibitors – prevent capsid formation.

Viroids & Prions

Viroids

Small, circular RNA molecules without a protein coat, infect plants, Potato famine in Ireland, Resemble introns.

Prions

Prions are "infectious proteins". They are normal body proteins that get converted into an alternate configuration by contact with other prion proteins. They have no DNA or RNA. The main protein involved in human and mammalian prion diseases is called "PrP".

Prion Diseases

It is causes neurons to rapidly degenerate examples: Mad cow disease (bovine spongiform encephalitis: BSE). People in New Guinea used to suffer from kuru, which they got from eating the brains of their enemies.

7. MICROBIAL GROUPS IN SOIL, MICROBIAL TRANSFORMATIONS OF CARBON, NITROGEN, PHOSPHORUS AND SULFUR, BIOLOGICAL NITROGEN FIXATION.

Introduction

Microorganisms are used routinely in engineered waste treatment systems such as sewage treatment plants. They are also of critical importance in the recovery process of natural environments degraded by human activities, such as in the self-purification of streams receiving sewage and runoff, and the natural attenuation of industrial contaminants leaked or spilled onto the soil. On the other hand, microorganisms have the potential to create substantial environmental problems. For example, they may deplete oxygen, generate unpleasant tastes and odors, clog equipment, and corrode pipes.

On this day, we consider the prokaryotic groups, bacteria and archaea. We also examine the eukaryotic groups containing single-celled organisms: protozoans, algae, fungi and slime molds, even though they also include many multicellular, macroscopic species. There is a wide range of diversity within the world of microorganisms in terms of survival strategy: where they find energy, how they grow, and what environments they prefer.

Energy Sources

The two major sources of energy are chemical oxidation and photosynthesis.

Carbon Sources

Since it is a major constituent of cell materials, all organisms need a source of carbon. Heterotrophs (including fungi, protozoans, and most bacteria) require organic carbon, whereas autotrophs (algae and some bacteria) consume inorganic carbon (carbon dioxide and bicarbonate).

Microbial cells are also commonly classified on the basis of environments they prefer. Several factors are generally considered, including the presence of oxygen, temperature, salt tolerance, and pH. Strict aerobes require oxygen; cells able to grow at very low oxygen levels may be referred to as microaerophilic. Facultative anaerobes can grow with or without oxygen. Anaerobic metabolism may be respiratory (using a variety of inorganic terminal electron acceptors such as nitrate, nitrite, ferric iron, sulfate, or carbon dioxide) or fermentative (using an organic terminal electron acceptor). Anoxic in the absence of oxygen, and thus is equivalent to anaerobic. The ability to utilize nitrate and/or nitrite as alternative terminal electron acceptors (denitrification). Microbes thrive under cold temperature conditions, ranging from below 0°C to the mid-teens. Organisms that prefer moderate temperatures are referred to as m. Their temperature preferences range from the mid-twenties to high-30s or low-40°C. A relatively few organisms, mainly bacteria, archaea, and fungi, prefer above 45 to 50°C and are called thermophiles. Some prokaryotic extremophiles are (temp. optimum above 80°C), a few even growing at above 100°C. Certain soil bacteria live symbiotically in the roots of legumes (N_2-fixing bacteria) -convert inert N_2 gas into amines. Mycorrhizal fungi -provide increased N, P, K, and antibiotics. Ruminant bacteria of cattle, sheep benefit these animals in metabolizing cellulose. Conversion of dinitrogen gas (N_2) to ammonia (NH_3). Availability of fixed N is the factor most limiting to plant growth. N-fixation ability is limited to few bacteria, either as free-living organisms or in symbiosis with higher plants. First attempt to increase forest growth through N-fixation in Lithuania, 1894 (lupines in Scots pine). Rare, extremely energy-consuming conversion because of stability of triply bonded N_2. Produces fixed N, which can be directly assimilated into N-containing biomolecules.

N-fixation requires energy input

Reduction reaction, e^- must be added (sensitive to O_2). Requires ~35 kJ of energy per mol of N fixed (theoretically). Actual cost: ~15-30g CH per g of NH_3 produced. Assimilation of NH_3 into organic form takes 3.1-3.6 g CH.

Rhizobia fix nitrogen in symbiotic association with leguminous plants. Rhizobia fix N for the plant and the plant provides Rhizobia with carbon substrates. All nitrogen-fixing systems appear to be identical. They require nitrogenase, a reductant (reduced ferredoxin), ATP, O-free conditions and regulatory controls (ADP inhibits and NH_4^+ inhibits expression of nif genes. Biological nitrogen fixation is the reduction of atmospheric nitrogen gas (N_2) to ammonium ions (NH_4^+) by the oxygen-sensitive enzyme, nitrogenase.

Reducing power is provided by NAPH/ferredoxin, via an Fe/Mo center. Plant genomes lack any genes encoding this enzyme, which occurs only in prokaryotes (bacteria). Even within the bacteria, only certain free-living bacteria (Klebsiella, Azospirillum, Azotobacter), blue-green bacteria (Anabaena) and a few symbiotic. Rhizobial species are known nitrogen-fixers. Another nitrogen-fixing association exists between an Actinomycete (Frankia spp.) and alder (Alnus spp.). The enzyme nitrogenase catalyzes the conversion of atmospheric, gaseous dinitrogen (N_2) and dihydrogen (H_2) to ammonia (NH_3), as shown in the chemical equation below: $N_2 + 3\ H_2 \Rightarrow 2\ NH_3$. The above reaction seems simple enough and the atmosphere is 78% N_2. The incredibly strong (triple) bond in N_2 makes this reaction very difficult to carry out efficiently. In fact, nitrogenase consumes ~16 moles of ATP for every molecule of N_2 it reduces to NH_3, which makes it one of the most energy-expensive processes known in nature Fig. 7.1.

Nitrogenase complex

Two protein components: nitrogenase reductase and nitrogenase: Nitrogenase reductase is a 60 kD homodimer with a single 4Fe-4S cluster. It is very oxygen-sensitive. It binds MgATP. The 4ATP required per pair of electrons transferred Reduction of N_2 to $2NH_3 + H_2$ requires 4 pairs of electrons, so 16 ATP are consumed per N_2. N_2 reduction to ammonia is thermodynamically favorable. However, the activation barrier for breaking the N-N triple bond is enormous and 16 ATP provides the needed activation energy. Each molecule of the enzyme contains 2 Mo, 32 Fe, 30 equivalents of acid-labile sulfide (FeS clusters, etc). Four 4Fe-4S clusters plus two FeMoCo, an iron-molybdenum cofactor. Nitrogenase is slow: - 12 e^- pairs per second, i.e., only three molecules of N_2 per second.

Types of N-fixers important in soils

Cyanobacteria: Autotrophic N-fixers, protect nitrogenase with specialized *heterocyst* cells. Heterotrophic bacteria: Free-living or associative with rhizosphere. Use energy from decomposing organic matter to fix N, protect nitrogenase by rapidly converting O_2 to CO_2 through respiration. Symbiotic bacteria: Plants form nodules to house bacteria and provide C as an energy source (*Rhizobium/Bradyrhizobium* for legumes, *Frankia* for non-legumes). Nodules contain a form of hemoglobin that binds O_2, protecting the nitrogenase enzyme.

Nitrogen fixation in Klebsiella

Nif system is turned on when No fixed nitrogen Anaerobic Temperature below 30°C. Nitrogenase is synthesized and converts N_2 to NH_3.

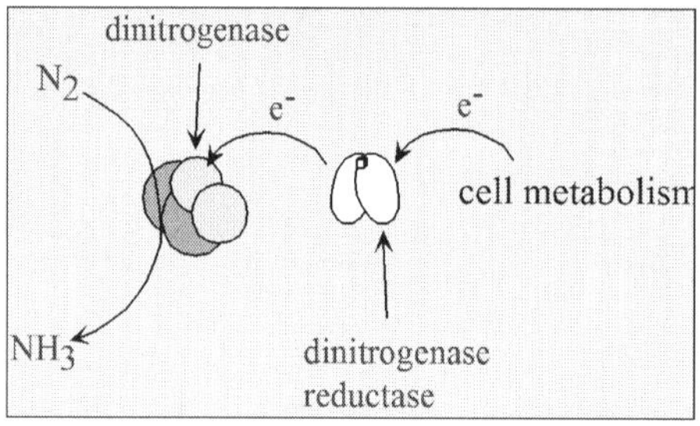

ADP ribosylation of dinitrogenase reductase

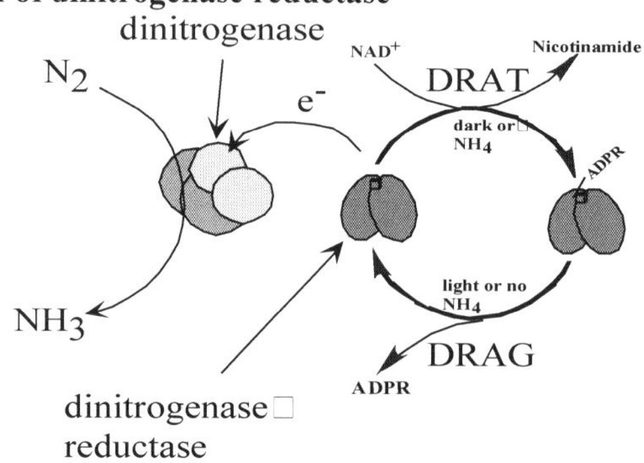

Bacterial associations with certain plant families, primarily legume species, make the largest single contribution to biological nitrogen fixation in the biosphere.

8. MICROBES IN COMPOSTING

Introduction

Composting is recycling naturally when we compost organic materials such as grass clippings, food and kitchen scraps, and leaves and brush, we are using the natural process of decay. Composting turns these wastes into a valuable humus-like product called compost. Composting is nothing more than speeding up the natural decay process. A compost pile is used to help us manage the "environmental factors that affect the rate of decay. These factors are air (more specifically oxygen) water, food, and temperature. The rate of composting (decay) and the quality of the resulting compost depend on how well you control these factors. Besides producing a valuable product, composting is recycling. It is one more way we can reduce the amount of waste we produce.

Pennsylvania is committed to recycling 35% of its municipal waste. Composting yard and kitchen wastes can help us maintain or exceed that recycling goal. Each one of us produces 650 lbs of waste each year that could be composted. Not too many people want another landfill next to their house, so why fill up the landfills we have with material that could be recycled. Not too many of us want higher bills for trash pick-up. By composting we can help keep down the cost of garbage handling. Over 60 percent of what we put in our landfills is organic waste that could be recycled by composting. Much of this can be recycled in your own backyard by composting.

Composting is a natural process and it is easy for you because you have a work crew that does almost all the work for you. Naturally occurring microorganisms (mainly bacteria and fungi) do the hard part of breaking down organic materials into compost. All you have to do to keep your decomposers working hard is provide them what they need. The right kinds of food, and The right amount of air, water, and warmth. While any organic material can be composted, you should not try to compost all of them in your backyard compost pile. Things that are good to compost in your backyard pile include: Garden trimmings, kitchen scraps, grass clippings, leaves, potting soil, manure, sawdust, hair and straw.

Some organic wastes are better left out of your backyard compost pile because they can cause problems either during composting or when you use the compost. These forbidden materials include: meat fish and dairy products because of their tendency to attract pests and vermin (flies, rodents). Hard to kill weeds such as bindweed and quackgrass as well as weeds that have produced seed. A hot compost pile can kill these weeds and seeds. Most backyard compost piles do not reach high enough temperatures for long enough periods of time to kill these weeds and seeds. Cat and dog manures also tend to attract pests and could spread disease. Again, most backyard compost piles do not get hot enough for long enough to be certain of destroying plant diseases or insects". So to be sure you don't spread or reinfect your garden, do not put diseased or insect-ridden plants into the compost pile.

Many people think it is necessary to add lime to the compost pile. "Finished compost is always near neutral in pH even if acidic materials were placed in the pile. For most garden uses, high pH compost is not desirable, and adding lime to the pile will often create an ammonia odor. Adding wood ash will add some potash to the pile which is good. But large amounts of wood ash will tend to increase the salt content of the finished compost. Many wood ashes are alkaline and will increase pile pH and lead to ammonia odor problems. Greens are nitrogen-rich materials, those that have carbon to nitrogen ratios that are less than 25 to 1. Greens also tend to be more wet and soft than browns. They decompose very rapidly.

Mixing compost into heavy soil improves its tilth and friability making it easier to work and easier for roots and water to penetrate. This also allows water to drain more readily so you don't have saturated conditions in your soil. Compost helps to loosen heavy clay soils and keep them from getting compacted and compost helps to suppress, or combat some soil microbes that can attack plant roots. Compost cannot be considered a fertilizer, but it does contain important plant nutrients such as nitrogen and phosphorus. Compost will slowly release these nutrients to plants. The nutrients in compost are not likely to be leached and lost from the topsoil. Compost also contains many trace nutrients that are essential to plant growth. Before using your compost be sure it is fully mature (use the bag test or germination test). Immature compost could damage or kill plants and could tie up soil nitrogen. When using compost as a soil amendment or soil conditioner, work no more than a 2-inch layer of compost into the upper 4 to 6 inches of soil. Compost used in this way will improve water and nutrient retention and will loosen heavy clay soils. Compost can be used as a surface mulch. Use a maximum depth of 3 inches for mulch. Never pile compost or any mulch around trunks of trees or shrubs. This makes an ideal environment for insects to attack the plant. Keep compost 3-4 inches away from the trunk and extend out to the dripline. Mulching protects the soil from temperature extremes and helps maintain soil moisture. Fully mature compost can be used as a topdressing on lawns. Used fine (screened) compost and apply about a ¼ inch layer to the lawn. The topdressing will be most effective if the lawn is cored before compost application. Rake the lawn after compost topdressing to move some compost into the core holes. Compost can replace peat in preparing potting mixes". The compost must also be very mature and fine (screened). Use no more than 1/3 compost by volume in your potting mix.

9. MICROBIOLOGY OF FOOD: MICROBIAL SPOILAGE AND PRINCIPLES OF FOOD PRESERVATION

Introduction

Various chemical and physical methods are available for the preservation of food. But the selection of the appropriate method is very important to protect the quality and nutritional value of the food. During preservation by these methods, the following principles are involved.

Microbial Decomposition of Food

Preventing contamination of food by spoilage organisms. "Removing micro-organisms from contaminated food. Forexample, by washing. Slowing down the growth and activity of micro-organisms in food. Forexample, by freezing the food. Killing micro-organisms present in the food for eg. by heating the food or by adding some chemicals.

Self-Decomposition of Food

Damaging self-enzyme of food that causes self-decomposition for eg. damaging of ripening enzyme of fruits by washing it in hot water. Preventing purely chemical reactions occurring in food that damage the food for eg. prevention of oxidative rancidity of lipid by addition of anti-oxidant. Prevention of mechanical damage of food caused by insects, birds, food handling devices ,etc.

Methods for Prevention of Microbial Spoilage of Food

1. Preventing food contamination

The process of preventing contamination of food by spoilage micro-organisms is called asepsis. Inner tissue of healthy animal and plants do not contain micro-organisms and if present they usually do not spoil food. Most spoilage micro-organisms enter into the food by contamination from external sources like water, air ,etc. Therefore, microbial decomposition of food can be prevented by preventing contamination of food by spoilage organisms. Several foods like egg contain natural covering around food that prevent the entry of spoilage organism into the food. Nowadays, a variety of such artificial barriers are made around the food to prevent contamination for eg. fruits and vegetables are stored for years by coating the wax around them.

Some other artificial barriers that prevent contamination include wrappers, cartons and cans. Safe handling of food and good personal hygiene of food handlers are important methods to prevent or minimize contamination of food by spoilage organisms.

2. Removing microorganisms from contaminated food

It is the second level of preservation of microbial decomposition of food. If contamination of food by spoilage organisms is not prevented, spoilage of food still can be prevented by removing contaminated micro-organisms from the food. Some other techniques that preserve food by removing micro-organisms include:

Filtration

It is only one successful method for removing micro-organisms from contaminated food. Unfortunately, this method is applicable only for clear liquids like clear fruit juice, beverages ,etc. To remove micro-organisms, liquid food is filtered through the sterile bacterial filter.

Centrifugation

Centrifugation removes some but not all micro-organisms from contaminated liquid food. When liquid food is centrifuged suspended particles, debris and some micro-organisms settle to the bottom which is later separated and remove from the food. This technique is commonly used for milk in dairy industries.

Washing

This method is commonly used for vegetables and fruits. Dusts and micro-organisms present on the surface of whole vegetables and fruits are removed by washing with sterile water. One limitation of washing is that it increases the water activity of some food and hence facilitates the growth of remaining spoilage organisms. Furthermore, if washing water is not sterile" it adds spoilage organisms in food.

Trimming

Trimming is the process of removing spoiled parts of food like vegetables and fruits by cutting. Trimming prevents the spread of micro-organisms from spoiled parts into inner healthy tissue.

3. Killing microorganisms

Different methods that preserve food either by killing micro-organisms or by slowing down their growth and activity includes:

- Preservation of a high temperature
- Preservation by low temperature
- Preservation by drying
- Preservation by radiation (irradiation)
- Chemical agents used in the preservation of food

Preservation of Food by High Temperature

Heat is the most commonly used method of food preservation. Heat kills spoilage of "microorganisms by denaturation of cytoplasmic protein and enzyme. Two types of heat i.e. dry heat and moist heat are used in the preservation of food. Moist heat is more effective in killing micro-organisms than dry heat. Heat causes the denaturation of protein and enzymes by breaking down their bonds. Breaking of bond occurs by two mechanisms i.e. hydrolysis and oxidation. Hydrolysis requires water but oxidation does not. Hydrolysis requires less energy (heat) for breaking of bond than oxidation. Therefore, if moisture is present in food, bonds of protein and enzymes are broken down by hydrolysis even at lower temperatures. This is the reason why moist heat can kill micro-organisms completely at the condition of 121.5°C for 15 min but dry heat requires 160°C for 1-2 hrs to kill the same organisms. Various factors affect the killing of micro-organisms in food by heat.

Time-temperature relationship. "Density of micro-organisms in food, if the density is high, a higher temperature is needed. Types of micro-organisms present in food, e.g. Spore are difficult to kill than vegetative cells. The growth stage of micro-organisms, e.g. micro-organisms present in the log phase are easily killed but micro-organisms present in the late log phase and stationary phase are somewhat heat resistant. Micro-organisms in dry food are difficult to kill than in moist food. If food is acidic or alkaline, micro-organisms present in the log phase are easily killed by heat. On the other hand, if food is neutral, high temperature is needed to kill the same organisms. Penetration of heat into the inner layer of food also affects the killing of micro-organisms. The heat penetration into the food is affected by various factors as given below: Nature of material by which container of food is made, affects heat penetration eg. heat penetrates easily into a smaller and slim can than through larger and wider can. In the case of open heating of food, agitation or stirring of food facilitates heat penetration.

Pasteurization

Pasteurization is a selective or gentle heating technique originally developed by Lewis Pasteur for milk and milk products. Nowadays, pasteurization is also used for the preservation of other beverages like beer, fruit juice ,etc. Pasteurization is a selective heating technique and it kills some but not all micro-organisms present in milk. If milk is contaminated by some disease-causing micro-organisms, milk cannot be heated at boiling temperature to kill these micro-organisms. It is because heating at high temperature brings several undesirable changes in milk eg.. it kills beneficial lactic acid bacteria present in milk. Similarly, casein protein precipitate, lactose caramelizes, the flavor of milk changes. Such milk becomes unsuitable for the production of milk products like ice cream, cheese. Similarly, beverages like beer and fruit juice cannot be heated at higher temperatures because it brings several undesirable changes in flavor, taste ,etc. In pasteurization, milk and beverages are heated at lower temperatures that kills spoilage or pathogenic organisms without bringing undesirable changes in the product. There are three techniques of pasteurization of milk which differ in their time-temperature relationship". Time-temperature relation of pasteurization are selected on the basis of three bacteria that are transmitted through milk (*Mycobacterium bovis, Coxiella buretii* and *Brucella abortus*). Low temperature holding (LTH) or Vat pasteurization: In this method, milk is heated at 62.8°C for 30 minutes. In this method, milk is heated at 71.7°C for 15 seconds. Ultra-pasteurization: In this method, milk is heated at about 137.8°C for 2 seconds.

Canning

Canning is the process of preservation of food by keeping them in a sealed container and then heating the container. Canning techniques were originally developed by Nicoles Appert. There are two types of canned food i.e. heated and unheated canned food. During canning at first raw food is washed and then processed. Then processed food is placed into a suitable can and the can is then heated. Sealed cans can be heated either by immersing it in boiling water or by a spray of super-heated steam or heated dry air.

Blanching or Scalding

Blanching is the process of washing vegetables or fruits in warm water before storage.

Advantages of Blanching Include:

- It inactivates the enzymes of food that cause self-decomposition.
- It reduces the number of micro-organisms present on the surface of the food.
- It brings wilting of leafy vegetables and helps in packaging.
- It enhances the fixing of the intense green color of fruits and vegetables.

Roasting

It is used for meat and meat products. During roasting internal temperature of meat reaches (80-85)°C.

Baking

It is used for bread, cake and other bakery products. During baking temperature maximum of 97°C is reached.

Frying

It is used for vegetables, meat, fish, etc. During frying outer surface of food becomes very hot but the internal temperature of food never exceeds 100°C.

Cooking

During cooking temperature around 100°C is reached.

Simmering

It is the process of gentle boiling of food with temperature of about 100°C.

Preservation of Food by Low Temperature

Low temperature is a commonly used physical method of preservation of food. "Low temperature preserves food by: slowing down growth and activity of spoilage organism in food. By slowing down the rate of a chemical reaction or other enzymatic activity that causes self-decomposition of food.

Sometimes low temperature kills spoilage organism in food eg. in case of deep freezing.

Different Methods of Food Preservation by Low Temperature

Common or Cellar Storage

Common storage is the technique of preservation of certain foods like potatoes, onion, etc. in the home by spreading them in cold rooms. In common storage, storage temperature is slightly lower than that of outside air. But this technique food is not stored for a long time but it is still used where cold storage and facilities are not available.

Chilling or Cold Storage

In chilling or cold storage, storage temperature is usually between (0-10) °C. At this temperature, the growth and activity of most bacteria except psychrophiles are significantly slowed down. Therefore, by chilling storage, food can be stored for a long time than common storage. However, if psychrophilic organisms are present in food, they grow in food and cause spoilage. Selection of appropriate temperature is very important because the quality of some food is damaged at a lower temperature storage temperature of some food are (10-12.8)°C for sweet potato, (13.3-16.7)°C for banana, etc. This method is most commonly used for short-term storage of highly perishable foods like meat, milk, fish, etc. Fish is stored by placing it in cold ice.

Freezing or Frozen Storage

In freezing storage, storage temperature is between (-15 to -29)°C. At this temperature, the growth of even psychrophilic bacteria is completely inhibited. Therefore, food kept in frozen storage remains fresh for a long time without undergoing microbial decomposition. Furthermore, self-decomposition of food occurs very rarely or not at all at this temperature. Sometimes, freezing also kills spoilage organisms present in food. Therefore, frozen storage is a very effective method of food preservation among these three techniques. The killing mechanism of freezing is due to the formation of ice-crystal and dehydration of microbial cells. When food is frozen, surrounding water freezes to form ice-crystals. Then cytoplasmic water from microbial cells continuously diffuses out and is added to surrounding ice-crystals. The remaining cytoplasm in microbial cells gradually becomes more and more concentrated. High salt concentration in the remaining cytoplasm kills the microbial cell by causing mechanical damage to the cell. There are two methods of freezing the food i.e. quick freezing and slow freezing. Slow freezing is more effective in killing microbial cells but quick freezing is considered more appropriate for the preservation of food. When food is frozen very slowly, there is sufficient time for cytoplasmic water to diffuse out. In this case microbial cell is dehydrated to a greater extent. Similarly, larger ice-crystal are formed due to the continuous addition of diffused water into ice. Larger ice crystal gives more mechanical damage to microbial cell". However, if larger ice-crystal are formed in food, it gives more mechanical damage (causes cracking of food) to the food. Therefore, quick freezing is commonly used for food preservation. Although frozen storage completely prevents microbial decomposition, it brings several undesirable changes in some food.

Some such examples include: Protein of food e.g. meat may become dehydrated so that the food becomes very tough and rigid. Myoglobin present in meat may be oxidized into brown-colored meat-myoglobin. Fat present in food may be hydrolyzed or oxidized. Ice-crystals formed on the surface of food evaporate and give mechanical damage to the surface of food. This damage is called freeze-burn.

Methods of Freezing

Quick Freezing

In quick freezing, the temperature of the food lowered very quickly. Food is frozen quickly by immersing food into the solution of refrigerant for eg. fish frozen by immersing it into a frozen solution of NaCl. By blowing frigid air (at -17.8 to -34.4°C) over the food to be frozen.

Slow Freezing

In this process, the temperature of the food is lowered very slowly. Food is frozen slowly by placing it inside the mechanical refrigerator. This technique is commonly used to freeze food in homes. However, slow freezing forms larger ice-crystal in food and causes more mechanical damage. Therefore, quick freezing is preferred over slow freezing.

10. BENEFICIAL MICROORGANISMS IN AGRICULTURE: BIOFERTILIZER (BACTERIAL CYANOBACTERIAL AND FUNGAL), MICROBIAL AGENTS FOR CONTROL OF PLANT DISEASES, BIODEGRADATION, BIOGAS PRODUCTION, BIODEGRADABLE PLASTICS

Generally, agricultural land gets impoverished after long-term cultivation, if not supplemented properly with inputs. To supplement the soil nutrient content under the conventional farming system, we need to apply high doses of agrochemicals which in turn pollute the ecosystem. "Therefore, in order to make agriculture sustainable, it is necessary to implement a balanced and responsible use of organic agriculture. The principles of organic farming also outline similar concepts where soil health and biodiversity are built up to sustain plant growth in the longer term. The name itself is self-explanatory. Biofertilizer is a ready-to-use live formulation of such beneficial microorganisms which on application to seed, root or soil, mobilize the availability of nutrients by their biological activity. They help build up the soil micro-flora and thereby the soil health. As we know, organic farming excludes the use of any chemical. The use of bio-fertilizer is recommended for improving soil fertility in organic farming. A simple form of classification of biofertilizers is given below:

Nitrogen

Rhizobium for legume crops. Azotobacter, Azospirillum for non-legume crops. Acetobacter for sugarcane only. Blue -Green Algae (BGA) and Azolla for low land paddy.

Phosphorous

Phosphatika for all crops to be applied with rhizobium, Azotobacter, Acetobacter. VAM (Vesicu1ar-arbuscular mycorrhiza).

Types of Biofertilizer

Nitrogen-fixing Biofertilizers

The nitrogen-fixing bacteria work under two conditions, symbiotically and as free-living bacteria (non-symbiotic). The symbiotic bacteria make an association with crop plants by forming nodules in their roots. The free-living bacteria do not form any association but live freely and fix atmospheric nitrogen. Now let us examine the features of these microbes in detail.

Rhizobium

Rhizobium lives in the root hairs of the legumes by forming nodules. The name *Rhizobium* was established by Frank in 1889. This genus has seven distinct species based on "Cross Inoculation Group Concept". More than twenty cross-inoculations groups have been established so far. A new classification has been established for *Rhizobium*" That is 'slow growing rhizobia' known as *Bradyrhizobium* and the other group is 'fast growing rhizobia' called *Rhizobium.* Still, this classification is discretely not distinguishable because the bacteria of one group may infect another group. This is called "the principle of cross inoculation" which relies on the assumption that legumes within a particular infection group may be nodulated by another species of nodule forming bacteria.

Rhizobium Inoculants

The seed treatment has been found to be the suitable method of "*Rhizobium* inoculation. Some adhesive is used to make proper contact between seeds and inoculants (bacteria). About 900 gm soil base culture is sufficient to inoculate the seeds for one-hectare area in the case of legumes. A 10 % jaggery *(gur)* solution is used as a base for *Rhizobium* cells to seed. First, the solution is spread over the seeds and mixed to build up a thin coat over the seeds. After ascertaining the proper coating of slurry over the seeds, the inoculant is sprinkled over the seeds and the content is again mixed thoroughly. Then the content is dried in the shade by spreading it thinly on a polythene sheet at least overnight. This is a free-living or non -symbiotic bacteria (does not form nodules but makes an association by living in the rhizosphere). Azospirillum species establish an association with many plants particularly with C4 plants such as maize, sorghum, sugarcane, etc. It is the most common organism and can form associative symbiosis with a large variety of plants. Azospirillum is recognized as a dominant soil microbe.

Blue-green Algae (BGA)

This is another important class of biofertilizer. The Blue-green algae are small organisms and can be seen under the microscope as a single cell or large accumulation of cells (colonies) or strings of cells (trichomes). They have a similar external appearance to that of algae and Azolla growing in a pond their requirements for light, nutrients and carbon dioxide are also similar.

Azotobactor

Azotobactor is a heterotrophic free-living nitrogen-fixing bacteria present in alkaline and neutral soils. *Azotobactor chrococcum* is the most commonly occurring species in arable soils of India. Apart from its ability to fix atmospheric nitrogen in soils, it can also synthesize growth promoting substances *viz.,* auxins, and gibberellins and also to some extent the vitamins. Many strains of *Azotobacter* also exhibit fungicidal properties against certain species of fungus. Response of *Azotobacter* has been seen in rice, maize, cotton, sugarcane, pearl millet, vegetable and some plantation crops. Field experiments have been carried out on *Azotobacter* .

Features of Azotobacter

Azotobacter contributes to the moderate benefits. *Azotobacter* is the heaviest breathing organism and requires a large amount of organic carbon for its growth. It is a poor competitor for nutrients in soil. It can benefit crops by nitrogen fixation, release of growth promoting substances, and fungicidic substances. *Azotobacter* is less effective in soils with poor organic matter content". It improves seed germination and plant growth.

Acetobactor

Acetobactor diazotrophicus is a newly discovered nitrogen-fixing bacteria associated with sugarcane crop. This bacterium belongs to the alpha group of *proteobacteria.* "It was isolated from leaf, root, bud and stem samples of sugarcane. Acetobator is located in the apoplastic fluid of sugarcane stems and to some extent in xylem vessels. It is an acid and high salt tolerant and sucrose loving bacteria that can fix up to 200 kg nitrogen per hectare.

Vesicular arbuscular mycorrhiza (VAM)

The term mycorrhiza was taken from Greek language meaning 'fungus root'. This term was coined by Frank in 1885. As indicated above, mycorrhiza is a mutualistic association between fungal mycelia and plant roots. VAM is an endotrophic (live inside) mycorrhiza formed by aseptated phycomycetous fungi.VAM help in nutrient transfer mainly of phosphorus, zinc and sulfur.

They also mobilize different nutrients like Cu (copper), K (potassium), Al (aluminum), Mn (manganese), Fe (iron) and Mg (magnesium) from the soil to the plant roots. They possess vesicles (sac-like structure) for storage of nutrients and arbuscular for funneling them into the root system.

Mechanism Of Action

The VAM forms an association with plant roots. It penetrates the root cortex and spreads around the roots of the plant. As the name indicates, they possess sac like structure called vesicules which stores phosphorus as phospholipids. The other structure called arbuscule helps bringing the distant nutrients to the vesicules and root.

Actions of Mycorrhiza

The feeding areas of the plant root are as the hyphae spread around the roots are enhancers to mobilizes the nutrients from distance to root. Stores the nutrients (phosphorus). It removes the toxic chemicals (example: phenolics) which otherwise hinder nutrient availability. It provides protection against other fungi and nematodes. This method is mostly used for fruit crops, sugarcane, and other crops where a localized application is needed. At the time of planting of fruit trees, 20 gm of biofertilizer mixed with compost is to be added in the ring of one sapling. Sometimes, the biofertilizers are also applied in the soil but we may require four to ten times more bio fertilizers. Before broadcasting, the inoculants should be incubated with the desired amount of well decomposed granulated FYM for 24 hours.

Advantages

There are many advantages of using the biofertilizers. They form an important association with other soil microbes and help in nutrient supply. Fixes atmospheric nitrogen Increase availability or uptake of nutrients through solubilization or increased absorption. Stimulate plant growth through hormonal or antibiotics action or by decomposing organic waste. They are cheap, hence, reduced cost of cultivation". Improves soil properties and sustaining soil fertility. Lead to soil enrichment. It is compatible with long-term sustainability. Build up soil fertility in the long term. They are eco-friendly and pose no damage to the environment.

Disadvantage

As such, there is no harmful impact of biofertilizers if it is used properly some constraints: specific to the plants. "*Rhizobium spp*. culture doesn't work well in high nitrate tolerant strains of soybean. The acceptability of biofertilizers has been rather low chiefly because they do not produce quick and spectacular responses. It is requiring skill in production and application. It is difficult to store.

Microbial Agents for Control of Plant Diseases

Rhizosphere microorganisms are found in the area surrounding and influenced by the plant roots and rhizoplane microorganisms are found on the surface of plant roots fig. 10.1.

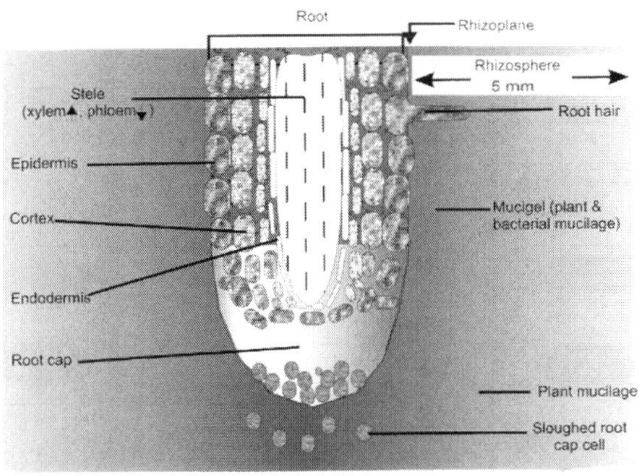

Figure 10.1: Rhizosphere.

Exudates: carbohydrates and proteins secreted by roots; Attracts bacteria, fungi, nematodes, protozoa. Bacteria and fungi are like little fertilizer bags. Nematodes and protozoa eat and excrete the fertilizer. Biological control agents are expensive and host-specific. Chemical pesticides are cost-effective, easy to apply and broad-spectrum.

Chemical Pesticides

Implicated in ecological, environmental, and human health problems. It is requiring yearly treatments and broad-spectrum. It is toxic to both beneficial and pathogenic species.

Biological Control Agents

Non-toxic to humans, not a water contaminant concerns and host-specific Only affect one or few species

Biological Control of Plant Pathogens

Antagonism

It is biological agents that reduce the numbers or disease-producing activities of the pathogen through one or more of the following mechanisms. "Antibiosis inhibition of one organism by another as a result of diffusion of a metabolic product such as antibiotic, lytic agents, enzymes, volatile compounds and other toxic substances. Antibiotic production is common in soil bacteria and fungi, for example: zwittermicin antibiotic production by *Bacillus cereus* against *Phytophthora* root rot in alfalfa. Nutrient competition between microorganisms for limited sources such as carbon, nitrogen, O_2, iron, and other nutrients or spaces. The most common way organisms are limit the growth of other: destructive mycoparasitism: t occurs when the antagonist invades the pathogen by secreting enzymes such as chitinase, cellulase, glucanase, and other lytic enzymes. The parasitism of one fungus by another direct contact, cell wall degrading enzymes, some produce antibiotics example *Trichoderma harzianum,* used as seed treatment against pathogenic fungus *Trichoderma spp.* are present in nearly all agricultural soils antifungal abilities have been known since the 1930s, Mycoparasitism, nutrient competition, agriculturally used as a biocontrol agent and as a plant growth promoter. Root colonization by *Trichoderma* strains frequently enhances: root growth and development. The crop productivity (increase up to 30%), resistance to biotic stress, uptake and use of nutrients. It is enhancing the solubilization of soil nutrients and increases root hair formation.

Biodegradation

Biodegradation – a biological activity that results in the breakdown of a specific contaminant.

Biodegradation

Bioavailability: bacteria are located at the oil-water interface, not inside the oil droplets. Smaller droplets give more surface area to increase biodegradation. Explosions in bacterial growth would deplete available nutrients and ultimately slow degradation. Oxygen consumption is possible dead zones. Sedimentation is deposition in anaerobic zones in sediment. Hazen *et al.* (2010) found 2-fold higher cell densities within the plume. Enrichment in plume for *Oceanospirillales*. Some oxygen depletion in the plume, not enough to create anoxic "dead zones". It is enough oxygen loss to indicate aerobic activity. Increased degradation genes in plume". Valentine *et al.* (2010) found oxygen depletion in the plume to be driven by ethane and propane metabolism.

Biogas Production

Biogas is generated by the activity of anaerobic bacteria. "Biogas is comprised of about 60% of methane, 40% of carbon dioxide, and a small amount of hydrogen sulfide, nitrogen, and hydrogen. The heating value of biogas is about 60% of natural gas and about 25% of propane. Biogas has corrosive nature and storage of biogas is not practical. The technology of biodigester is widely used in developing countries such as China, Vietnam, India, and Central and South America as well as in developed countries. The anaerobic digester can be used in remote farm areas to produce biogas from manure and protect water resources. The idea is to make a small-scale, low-cost biodigester plant so that anyone in the world can make it and produce biogas. Feed the bag with the effluent. Cut the garden hose to an adequate length and pull the bag through the garden hose. It is folded against the surface of the garden hose. It is attached to the adapter to the garden hose over the bag. It is connecting the hose and the gas collector using adequate adapters.

Operation

The generator requires little maintenance besides occasional stirring when the batch is done, disconnect the garden adapter and unload the effluent. For the next batch, apply about 10% of the previous batch to activate the new batch. After loading the feedstock, connect the adapter again and repeat the process.

11. PLANT– MICROBE INTERACTIONS

Introduction

Heavy metals are the most important inorganic pollutants, which are not degraded and progressively accumulate in the environment. The use of plants for rehabilitation of heavy-metal-contaminated soils is an emerging area of interest because it provides an ecologically and environmentally sound and safe method for restoration and remediation. Although a number of plant species are capable of hyper-accumulation of heavy metals, however, this approach is not applicable for remediating sites with multiple contaminants. Biogeochemical capacities of microorganisms seem almost limitless and they can adsorb and accumulate metals in their cells and are being used in microbial leaching and also as agents of cleaning the environment. To overcome the metal stress, a number of mechanisms have been evolved by microorganisms of agronomic importance by which they tolerate and promote the uptake of heavy metal ions. The best approach would be to combine the advantages of plant-microbe interactions within the plant rhizosphere into an effective cleanup technology.

The activities of plants and plant-microbial associations may offer viable means of accomplishing the *in situ* remediation of contaminated soils. This section examines the potential role of plant-microbe interactions in heavy metal contaminated soils towards phyto-bioremediation. Heavy metal pollution affects the production and quality of crops, the quality of atmosphere and water bodies and thus threatens human and animal health (Narula et. 2011). The metal species commonly found in the soils as a result of human activities include, copper (Cu), zinc (Zn), nickel (Ni), lead (Pb), cadmium (Cd), cobalt (Co), mercury (Hg), chromium (Cr) and arsenic (As) ,etc. Some of these act as micronutrients at small concentrations for living organisms for their normal physiological activities, but their accumulation is toxic to most life forms (Khan et al. 2005). The most common human activities resulting in the entry of heavy metal into land are the disposal of industrial effluents, disposal of waste such as sewage sludge, atmospheric deposition from industrial activities, mining activities, domestic and industrial wastes, land fill operations and use of agrochemicals. The release of heavy metals from various industrial sources, agrochemicals and sewage sludge present a major threat to the soil environment. Generally, heavy metals are not degraded biologically and persist in the environment indefinitely (Walker et al. 2003).

The toxic heavy metals inversely affect the microbial compositions, including plant growth promoting rhizobacteria (PGPR) in the rhizosphere, and their metabolic activities. In addition, the elevated concentration of metals in soils and their uptake by plants adversely affect the growth, symbiosis and consequently the yields of crops (Wani et al. 2008a) by disintegrating cell organelles, and disrupting the membranes (Sresty and Madhava Rao 1999), acting as genotoxic substance disrupting the physiological process, such as photosynthesis (Wani et al. 2007b) or by inactivating the respiration, protein synthesis and carbohydrate metabolism (Shakolnik 1984). The remediation of metal-contaminated soils thus becomes important, as these are rendered unsuitable for sustainable agriculture. This review provides a short overview on plant-microbe interactions towards phyto- and bio-remediation.

The phytoremediation approach involves the cultivation of metal accumulating higher plants to remove contaminants from metal-polluted soils (Brooks 1998). In this approach, plants capable of accumulating high levels of metals are grown in contaminated soils. At development metal enriched above ground biomass is harvested and soil metal contamination is detached. Successful plant-based decontamination of even moderately contaminated soils would need crops able to concentrate metals in excess of 1-2%. Accumulation of such high levels of heavy metals is highly toxic and would certainly kill the common nonaccumulator plant. However, in hyper-accumulator species, such concentrations are attainable. Nevertheless, the extent of metal removal is ultimately limited by the plant's ability to extract and tolerate only a finite amount of metals. On a dry weight basis, this threshold is around 3% for Zn and Ni, and considerably less for more toxic metals, such as Cd and Pb. The other biological parameter which limits the potential for metal phytoextraction is biomass production. With highly productive species, the potential for biomass production is about 100 tons' fresh weight/hectare. The values of these parameters limit the annual removal potential to a maximum of 400 kg metal/ha/yr. It should be mentioned, however, that most metal hyper-accumulators are slow-growing and produce little biomass. These characteristics severely limit the use of hyper-accumulator plants for environmental cleanup.

Practices have been residential to increase the potential of common non-accumulator plants for Pb phytoextraction. Particularly, the uptake-inducing properties of synthetic chelates open the possibility of using high biomass-producing crops for Pb phytoextraction. Under chelate-induced conditions, maize (Huang and Cunningham, 1996) and Indian mustard (Blaylock et al., 1997) have been successfully used to remove Pb from solution culture and contaminated soil, respectively. Physical characteristics of soil contamination are also important for the selection of remediating plants. For example, for the remediation of surface-contaminated soils, shallow rooted species would be appropriate to use, whereas deep-rooted plants would be the choice for more deep contamination.

The identification of metal hyperaccumulators, plants capable of accumulating high metal levels, demonstrate that some plants have the genetic potential to clean up metal contaminated soils. In general, the concentration of metals in hyperaccumulators is about 10-100 fold higher than most other plants growing on metal contaminated soils. It has been possible through bioengineering to develop plants (Raskin 1996) capable of removing methyl mercury from the contaminated soil. To detoxify this compound, such bioengineered plants express modified bacterial genes *merB* and *merA* which convert methyl mercury to elemental mercury. About 400 plant species have been identified as hyper-accumulator accumulators. The Indian mustard plant (*Brassica juncea*) can extract both heavy metals and radionuclides from the soil. Panwar et al. (2002) reported that *B. juncea* has the potential to be a hyper-accumulator of Ni. A Survey of literature reveals that the rate of metal removal depends upon the plant species, soil contaminating heavy metal(s), biomass harvested and metal concentration in harvested biomass as summarized in the table 11.1 below.

Table 11.1. Plants capable of hyperaccumulating metals

Plant species	Metal accumulated	Accumulated concentration mg/kg dry matter
Thlaspi caerulescens **(Brassicaceae)**	Zn	**10,000**
Seberatia acuminate **(Sapotaceae)**	Ni	**10,000**
Alyssum lesbiacum **(Brassicaceae)**	Ni	**20,000**
Arbidopsis halleri **(Brassicaceae)**	Cd	**1,000**
Thlaspi rotundifolium	Pb	**8,200**
Astralagus **sp.(leguminosae)**	Se	**1,000**
Pteris vittata **(Fern)**	**As**	**22,630**

Rhizosphere

The rhizosphere is the zone of soil surrounding a plant root where the biology and chemistry of the soil is influenced by the roots, rather it is an area of intense activity (biological, chemical and physical) influenced by compounds exuded by roots and by micro-organisms feeding on these compounds (Kumar et al. 2007). Generally, soil contains bacterial numbers in the range of 10^7 to 10^{10} cells per gm dry soil. But microbiological activity in the rhizosphere is much greater (10^8) than in soil away from plant roots (10^5) and also microorganisms provide or make available nutrients for the plants (Walker et al. 2003). Many of these microbes live there as a part of a distinct community surrounding plant roots. Heterotrophic bacteria are able to use organic compounds excreted in root exudates, whereas their metabolites can be used by other microbes, which in the end creates a network of closely associated microorganisms. This phenomenon of highly active micro-organisms in root-associated soil is known as the "rhizosphere effect". Thus the microbial population is one of the essential parts of the rhizosphere that affect the rhizosphere soil by its various activities such as water and nutrient uptake, exudation, and all the biological transformations. Among the fast-growing and early colonizing bacteria attracted by the plant exudates are members of genera *Bacillus* and *Pseudomonas* besides N fixing bacteria *Azospirillum* and *Rhizobium*.

Cultivation-based methods show that the *Pseudomonas* spp are generally more abundant in the rhizosphere than in the bulk soil. But the number of related clones to *Pseudomonas* is higher in the rhizosphere of ryegrass and white clover. On the basis of 16s rRNA gene clones, plant roots have further been shown to have a selective effect towards r-proteobacteria leading to the majority of *Pseudomonas* spp in rhizosphere as compared to bulk soil. Free-living and associative diazotrophs are more abundant in the rhizosphere than in bulk soils, indicating their dependence on organic compounds exuded by roots in the rhizosphere. Various nitrogen-fixing micro-organisms have been found to be present in the rhizosphere of agricultural plants, but the contribution of fixed nitrogen to plant nutrition is controversial (Lima et al. 2006). Diazotrophs found in the soil or associated with roots include *Azotobacter chroococcum*, *Azospirillum brasilense* and *Gluconacetobacter diazotrophicus* (formerly *Acetobacter diazotrophicus*) and the positive responses of plants to inoculation with these bacteria are attributed to nitrogen (N_2) fixation besides, several other factors like phytohormone and/or ammonium production" etc. (Okon 1985). Plant-associated rhizobacteria and mycorrhizae may significantly increase the bioavailability of various heavy metal ions for their uptake by plants. Also, they are known to catalyze redox transformations leading to changes in heavy-metal bioavailability (Yang et al. 2005).

Plant association with diazotrophs or any colonizer indicates a high degree of adaptation between the host plant and the most abundant diazotrophs. The micro habitat provided by the host plant seems to generate a selection pressure in favor of the micro-organisms, which in turn best benefit the host. Nutrients and metals are typically present in the soil solution at low concentrations and tend to form sparingly soluble minerals (except nitrogen, sulfur, and boron), or may be adsorbed to a solid phase through ion exchange, hydrogen bonding, or complexation (White 2003). The extent to which they are transferred from the soil to the biota (i.e., microbes or plants) is dependent on the biogeochemical interactions (N, P, S) and/or processes among the soil, plant roots, and microorganisms in the rhizosphere (Abbot and Murphy, 2003). At this interface, the presence of root exudates may influence chemical reaction kinetics within the soil environment and subsequently affect biological activities. The exudates act as messengers that "stimulate biological and physical interactions between root and soil organisms. In addition to the adsorption and conduction, roots also produce hormones and other substances that help to regulate the plant's development and structure, help in modifying the biochemical and physical properties of the rhizosphere (Abbot and Murphy 2003) and contribute to root growth and plant survival.

PGPR (Plant Growth Promoting Rhizobacteria)

Plant growth promoting rhizobacteria (PGPR) are capable of promoting plant growth by colonizing and establishing around the plant root (Narula et al. 2006). "PGPRs have been found to play a potential role in developing sustainable systems in crop production (Shoebitz et al. 2009). Soil bacteria have been used as biofertilizers for ameliorating soil fertility and enhancing crop production for decades. The main functions of these bacteria are (1) to supply nutrients to crops; (2) to stimulate plant growth, e.g., through the production of plant hormones; (3) to control or inhibit the activity of plant pathogens; (4) to improve soil structure; and (5) to act as bio-accumulator in microbial leaching of inorganics (Ehrlich. 1990). More recently, bacteria have also been used in soil for the mineralization of organic pollutants, i.e. bioremediation of polluted soils (Zaidi et al. 2008). Generally, PGPRs function in three different ways (Glick 1995, 2001): a) synthesizing particular compounds for the plants (Dobbelaere et al. 2003), b) facilitating the uptake of certain nutrients from the soil (Çakmakçi et al. 2006) and c) lessening or preventing the plants from diseases. Besides their role in protecting the plants from metal toxicity, the plant growth promoting rhizobacteria are also known for their role in enhancing soil fertility and promoting crop productivity by providing essential nutrients (Zaidi and Khan. 2006) and plant growth regulators (Kumar et al. 2007).

Glick et al. (2002) reported that these PGPRs also promote the growth of plants by alleviating the stress induced by ethylene-mediated impact on plants by synthesizing 1-aminocyclopropane-1-carboxylate (ACC) deaminase (Belimov et al. 2005). According to their relationship with the plants, PGPRs can be divided into two groups (Khan 2005). The first group "Symbiotic bacteria" with species of *Rhizobium* (*Rhizobium, Mesorhizobium, Bradyrhizobium, Azorhizobium, Allorhizobium* and *Sinorhizobium*) have been successfully used worldwide to permit an effective establishment of the nitrogen-fixing symbiosis with leguminous crop plants (Bottomley and Maggard 1990; Lynch 1990). The second group "Free-living rhizobacteria" with non-symbiotic nitrogen-fixing bacteria such as *Azotobacter, Azospirillum, Bacillus* and *Klebsiella* sp. are also used to inoculate a large area of arable land in the world with the aim of enhancing plant productivity (Lynch 1983). In addition, phosphate solubilizing bacteria such as species of *Bacillus* and *Paenibacillus* (formerly *Bacillus*) have been applied to soils to specifically enhance the phosphorus status of plants.

The mechanisms have been evolved by microorganisms of agronomic importance by which they tolerate and promote the uptake of heavy metal ions. Such mechanisms include (1) the pumping of metal ions exterior to the cell (2) accumulation and sequestration of the metal ions inside the cell (3) transformation of toxic metal to less toxic forms (Wani et al. 2008b) and adsorption/desorption of metals (Mamaril et al. 1997). Due to these properties, when plant growth promoting rhizobacteria including nitrogen fixers, used as seed inoculants (Narula et al. 2005a, 2005b), were applied to soil, either treated/amended intentionally with metals or already contaminated, have shown a substantial reduction in the toxicity of metals and concomitantly improved the overall growth and yield of chickpea (*Cicer arietinum*) (Gupta et al. 2004), green gram (*Vigna radiata* L. wilczek) and pea (*Pisum sativum*) (Wani et al. 2007a, 2007c). An experiment on the effect of rhizobial inoculation on the metal accumulation potential of the chickpea plant was done by Gupta et al. (2004). The chickpea plants were harvested and accumulation of metal was studied in root and shoot parts of the plant. Accumulation of Cu, Zn, Cr and Cd was more in shoot whereas Fe was more in the root part of the plant. These studies are significant while using the plant for metal decontamination of fly ash land fills. The use of such microbes possessing multiple properties like plasmid-borne (Saitia et al.

1989) metal resistance/reduction (Saitia and Narula 1989, Narula et al. 2011) and their ability to promote plant growth through different mechanisms in metal-contaminated soils are chosen as a suitable choice for bioremediation studies.

In the era of sustainable crop production, the plant-microbe interactions in the rhizosphere play an important role in the transformation, mobilization, solubilization, etc. of nutrients from a limited nutrient pool, and subsequent uptake of essential nutrients by plants to realize their full genetic potential. An understanding of the mechanisms, which is important for the initiation and establishment of the association between host and bacterium, can be reached from the analyses of influences exerted by each interaction partner on the other. In addition, there is a need to know how diazotrophs may benefit the plant. Investigations on the production of phytohormones and the action of siderophores produced by *Azotobacter* strains might help to understand this aspect of interactions (Narula et al. 2006; Narula et al. 2005a, 2005b). Plant-microbe interactions are important for both the partners i.e., macro as higher plants and micro partners as the plant-associated bacteria (Somers et al. 2004). Microbial partners can induce antagonistic (in the case of phytopathogens) or symbiotic interactions. Different types of interactions involving plants roots in the rhizosphere have been reviewed by Bais et al. (2004, 2006). These include root-root, root-insect, and root-microbe interactions. The rhizosphere represents a highly dynamic front to study the interaction between roots and pathogenic as well as beneficial soil microbes, invertebrates, and root systems of competitors (Bais et al. 2004). In recent years several plant scientists have recognized the importance of root exudates in mediating these biological interactions. However, because plant roots are always hidden below ground, many of the interesting phenomena, their attractions, love and hate relationship in which they are involved have remained largely unnoticed. Especially the role of chemical signals (Peters et al. 1986) in mediating below ground interactions is only beginning to be understood. Chemical signaling between plant roots and other soil organisms, including the roots of neighboring plants, is often based on chemicals exuded from the roots. The same chemical signals may elicit dissimilar responses from different recipients. Most importantly, chemical components of root exudates may be the ones to deter one organism while attracting another, or two very different organisms may be attracted with differing consequences to the plant.

Therefore, the mechanisms used by roots to communicate and interpret these signals which they receive from other roots and soil microbes in the rhizosphere are largely unknown. The rhizosphere has some positive or negative and neutral associations. Much has to be done still to determine and elucidate whether the chemical signature of a plant root exudates will be perceived as a negative or a positive signal. However, evidence over the years suggest that root exudates are the determining factor to identify interactions in the rhizosphere and, ultimately, plant and soil community dynamics (Narula et al. 2009).

Biotic and Chemical Interferences of Heavy Metals

In the last decades, the essential role of soil micro-biota has been emphasized for many processes that are directly related to soil bioremediation and, especially, phytoremediation, in which plant-microbe interactions have been documented to be of vital importance (Khan 2005). Phytoremediation is a process that uses actively growing plant roots to stimulate a diverse population of soil microorganisms, some of which have the capability to metabolize contaminants. In the case of phyto-extraction, inoculation of plants with PGPR may be beneficial mainly due to plant growth stimulation and a corresponding increase in the above-ground biomass, rather than to an increase in metal concentrations in plant tissues, which may actually be weakly influenced (Wu et al. 2006). Such highly persistent and hardly avoidable pollutants as heavy metals, in contrast to organic contaminations (Karthikeyan and Kulakow 2003), cannot be biologically or chemically degraded, but can only be deactivated by detoxification and/or immobilization) within the soil or removed from the contaminated site (Ruml and Kotrba 2003).

On the other hand, they can directly or indirectly affect plant-microbe interactions in the rhizosphere. Their direct influence involves possible toxic or other detrimental effects of particular chemical metal species on the biochemistry and physiology of plant-associated microorganisms (Renella et al. 2005) and on the growth and development of plants and these effects comprise the biotic interferences. The chemical reactivity of particular metal species or their catalytic effects, which result in chemical binding or degradation of biogenic organics involved in plant-microbe interactions (i.e. remote exchange of molecular signals, quorum sensing within bacterial communities, release and exchange of nutrients, etc. (Gunatilaka 2006), may thus be classified as indirect effects comprising chemical interferences". Biotic processes, induced in the rhizosphere by bio-geo-chemical interactions or formation of chemically active metal species, results in chemical depletion, inactivation or degradation of any biomolecules directly involved in plant-microbe interactions *via* their binding and/or by redox transformation, invariably affect these biologically specific interactions as described by various researchers (Lasat 2000; Kamnev 2008).

Metal Bioavailability

A metal uptake into roots is slow transport from soil particles to root surfaces. There are two mechanisms (Barber 1984) responsible for metal transport from the bulk soil to plant roots: 1) convection or mass flow, and 2) diffusion. Due to convection, soluble metal ions move from soil solids to the root surface. From the rhizosphere, water is absorbed by roots to replace water transpired by leaves. Water uptake from the rhizosphere creates a hydraulic gradient directed from the bulk soil to the root surface. Some ions are absorbed by roots faster than the rate of supply *via* mass flow. Thus, a depleted zone is created in soil immediately adjacent to the root. This generates a concentration gradient directed from the bulk soil solution and soil particles holding the adsorbed elements to the solution in contact with the root surface. This concentration gradient drives the diffusion of ions toward the depleted layer surrounding the roots. Plants have evolved specialized mechanisms to increase the concentration of metal ions in soil solution. For example, at low ion supply, plants may alter the chemical environment of the rhizosphere to stimulate the desorption of ions from soil solids into solution. Such a mechanism comprises rhizosphere acidification due to H^+ extrusion from roots (Crowley et al. 1991). Protons compete with and replace metal ions from binding sites, stimulating their desorption from soil solids into solution. In addition, some plants can regulate metal solubility in the rhizosphere by exuding a variety of organic compounds from roots. Root exudates form a complex with metal ions keeping them in solution available for uptake into roots (Romheld and Marschner 1986).

Role Of PGPR in Biotic Interferences of Heavy Metals

In cooperation microorganisms and plants have developed diverse strategies helping them to overcome or to adapt to unfavorable environmental conditions. A number of PGPR and other microorganisms have been reported to be resistant to relatively high concentrations of heavy metals (Ruggiero et al. 2005) and remain active in moderately acidic soils (Yasmin et al. 2004) which are wide-spread, comprising over 30% of only arable territories (Von Uexkull and Mutert 1995). Such phytostimulating rhizobacteria could contribute to plant growth promotion in metal-contaminated sites both indirectly, by increasing the overall fertility of the contaminated soil, as well as providing nutrients, phytohormones and exhibiting other PGPR-related traits (Rosenblueth and Martinez-Romero 2006) but also catabolising certain organics and/or partly oxidized intermediate biodegradation products.

As for the PGPR of the genus *Azospirillum*, studies have shown that these phyto-stimulating micro-aerophilic rhizobacteria are relatively tolerant to sub millimolar concentrations of heavy metals at which, in many cases, growth of the bacterial culture is not significantly suppressed (Kamnev et al. 2006). Moderate concentrations of heavy metals in soil can often be found as a result of contamination and/or bioleaching of minerals, and assessing their impact on the bacterial metabolism is of importance for a deeper insight into their biology and presents an obvious biotechnological and agricultural interest, considering the plant-growth-promoting abilities of Azospirilla (Somers et al. 2004). In particular, *A. brasilense* represents a relatively rare chance to compare two strains of the same species which essentially differ in their mode of plant-root colonization. For example, *A. brasilense* strain Sp245 is known to be a facultative endophyte, whereas non-endophytic strain Sp7 colonizes the root surface only (Kirchhof et al. 1997). The levels of the auxin phytohormone (indole-3-acetic acid, IAA) production for each of the two strains were analyzed in a standard medium (control) and in the presence of 0.2 mM $CuSO_4$ or $CdCl_2$ under micro-aerobic conditions that are favorable for these micro-aerophilic bacteria. In parallel, the bacterial growth rate (in terms of colony-forming units, CFU s) was also assessed. Using strain Sp7, both copper (II) and cadmium ions were found to result in a significant decrease in the level of IAA production, whereas the bacterial growth rate was virtually not affected. In Sp245, the overall level of IAA production in the control was approximately 3 times higher than for Sp7. Nevertheless, in this respect, strain Sp245 appeared to be more sensitive to heavy metals: a noticeable decrease in IAA production was observed under the effect of both the metals, especially with Cu (II); note also that for strain Sp245 copper (II) was somewhat more toxic decreasing also the bacterial counts.

Thus, the decrease in IAA production by the bacteria in presence of copper (II) and cadmium (II) may directly affect the efficiency of associative plant-bacterial symbioses in heavy metal-polluted soils. To compare heavy-metal resistance of some plant-associated soil bacteria, the values of minimum inhibitory concentrations (MIC) and minimal lethal concentrations (MLC) for several PGPR grown on the agar medium (aerobic conditions) have been reported by Tugarova et al. (2006). MLC of cadmium (II) found for both the two *A. brasilense* strains (5 mM) was much higher than those reported for the other associative rhizobacteria (0.35 to 0.6 mM), i.e. by an order of magnitude, while for the Cd-sensitive rhizobacterium *Flavobacterium* sp. L30 the MLC was 0.01 mM (Belimov et al. 2005). By their cadmium resistance, the following soil bacteria isolated from the rhizoplane of Indian mustard (*Brassica juncea* L. Czern.) plants (Belimov et al. 2005) were closer to *A. brasilense*: *Variovorax paradoxus* (MLC = 0.6 to 3.5 mM for different strains), a *Rhodococcus* sp., a *Ralstonia* sp., a *Flavobacterium* sp., a *Pseudomonas* sp. and some other strains (MLC = 2.0 to 4.0 mM). Soil microorganisms can also directly influence metal solubility by altering their chemical properties. For example, a strain of *Pseudomonas maltophilia* was shown to reduce the mobile and toxic Cr (VI) to nontoxic and immobile Cr (III), and also to minimize environmental mobility of other toxic ions such as Hg (II), Pb (II), and Cd (II) (Blake et al. 1993; Park et al. 1999). In addition, it has been estimated that microbial reduction of Hg generates a significant fraction of global atmospheric Hg 0 emissions (Keating et al. 1997).

Simon et al. (2008) reported that pre-treatment of Indian mustard with *Pseudomonas fluorescens* before cadmium and nickel application slightly enhanced the Cd or Ni accumulation in roots. This enhancement was not observed when pseudomonads were applied simultaneously with metals. More cadmium was detected in the roots of Indian mustard when plants were pre-treated with Cd-tolerant than Cd-sensitive *Pseudomonas cepacia*. The cell number of pseudomonads in the surface of Cd-treated Indian mustard roots was significantly lower when plants were pre-treated with Cd-sensitive *P. cepacia*. Only for *Bacillus* and *Pseudomonas* species present in Se contaminated soils can be supposed to stimulate the Se phytoextraction and/or phytovolatization in the rhizosphere of higher plants (Simon et al. 2007). Pal et al. (2010) reported the influence of Pb and Hg on *Azotobacter spp.* and *Pseudomonas spp.* multiplication using Indian Mustard as a test crop on an artificially spiked sandy loam soil. The results showed that the minimum bacterial counts were observed in Pb_{200} and Hg_{40} treatments at 54 days after sowing of the crop.

In turn, rhizospheric microorganisms may interact symbiotically with roots to enhance the potential for metal uptake. Root growth affects the properties of the rhizospheric soil and stimulates the growth of the microbial consortium. In addition, some microorganisms may excrete organic compounds (Narula et al. 2009) which increase bioavailability and facilitate root absorption of essential metals, such as Fe (Crowley et al. 1991) and

Mn (Barber and Lee 1974), as well as nonessential metals, such as Cd (Salt et al. 1995). The rhizosphere effect is often expressed quantitatively as the ratio of the number of microorganisms in rhizosphere soil to the number of microorganisms in non-rhizosphere soils (R/S ratio). This type of increased microbial activity in the rhizosphere may be responsible for the increased metabolic degradation rate of certain xenobiotic compounds in the rhizosphere. The actual composition of the microbial community in the rhizosphere is dependent on root type; plant species, plant age and soil type as well as other factors. The bacteria clearly showed a preference for one exudate over the other. This might well be related to the substances found in each exudate. It might be argued that the energy yield from C_6-compounds is better, thus resulting in observed preferences (Kumar et al. 2007). Generally, the rhizosphere is colonized by gram-negative bacteria. The interaction between plants and microbial communities in the rhizosphere is complex and has evolved to the mutual benefit of both organisms. Plants sustain a large microbial population in the rhizosphere by secreting substances such as carbohydrates and amino acids through root cells and by sloughing root epidermal cells. The magnitude of rhizodeposition can be quite large. In addition, root cells secrete mucigel, gelatinous substances that are lubricants for root penetration through the soil during growth. This mucigel along with other cell secretions consists of root exudates. Beneficial effects of the root exudates are stimulated by adding organic acids to the soil. This enhances the microbial population and the evolution of greater amounts of carbon dioxide. Microbial counts are generally more in the soil from vegetation and plants, this explains the help given by root exudation. Microbial activity, biomass and degradation are greater than in the non-vegetated soil.

Chemical Interferences

Microorganisms have been shown to alter metal bioavailability. For example, microbes have been documented to catalyze redox reactions leading to changes in metal mobility in soil. A strain of *Xanthomonas maltophyla* was shown to catalyze the reduction and precipitation of highly mobile Cr (VI) to Cr (III), a significantly less mobile and environmentally less hazardous component. Ionic Hg (II) which is susceptible to methylation and subsequent volatilization generally of toxic methyl mercury was shown to be reduced by the same bacteria to elemental Hg which possess less environmental risk. A heavy metal resistant *Bacillus circulans* have been isolated from soil which is capable of biosorption of 25, 22, 20, 13, 12 mg/l of Mn, Zn, Cu, Ni and Co, respectively.

Besides the biotic reduction of iron (III) mediated by specific soil bacteria dissimilatory reducers, a range of secondary metabolites produced both by soil microorganisms and by plant roots can abiotically reduce Fe^{III} (Kovacs et al. 2006). Another rapidly developing field within the "underground hemisphere" of plant-microbe interactions, that can be affected by chemical interferences in soil, includes biosynthesis and exudation of intercellular bacterial signaling (quorum- sensing molecules), as well as a variety of already known, undisclosed biomolecules by which plant root-microbe signaling and interactions are implemented (Sugawara et al. 2006) or by which plants mimic, quench or otherwise interfere with bacterial signals (Teplitski et al. 2004; Adonizio et al. 2006). In particular, noteworthy is the active role of PGPR cell-surface polysaccharide-containing macromolecular complexes (Somers et al. 2004) and cell-surface lectins, in bacterial interactions with host-plant roots involving also plant lectins. As an example, wheat lectin (wheat germ agglutinin, WGA), which is exposed on the root surface and can also dissolve there from in the aqueous medium, has been documented to induce a number of metabolic responses in Azospirilla at low concentrations and therefore is proposed to act as a molecular signal for these wheat-associated bacteria (Narula et al. 2011, in press).

Soil micro-flora is an important and active participant in biogeochemical processes, leading to diverse chemical transformations of soil minerals. In this respect, the rhizospheric microbial communities supported by host plants (along with a contribution from plant-root exudates) can demonstrate even higher activities (Mukerji et al. 2006). "Rhizosphere, as compared to the bulk soil, is highly populated by various microorganisms mainly comprising bacteria (predominating Gram-negative bacteria) and mycorrhizal fungi (Karthikeyan and Kulakow 2003; Lucy et al. 2004) showing higher metabolic activity even in polluted soils (Biró et al. 2006). Some of the diverse routes leading to microbially mediated bioleaching and biotransformation of heavy metals and radionuclide entrapped within soil minerals (Zachara et al. 2001; Kamnev et al. 2006) are: (i) microbial dissimilatory reduction and concomitant solubilization of metal ions (e.g., Fe, Mn and many other less abundant redox-active metals) *via* different mechanisms as reviewed by (Lovley and Coates 2000). (ii) siderophores, carboxylic acids, acidic exopolysaccharides and other organics production and excretion capable of binding and/or reducing metal cations (Faraldo-Gomez and Sansom 2003) as well as (iii) contact interactions of microbial cells with metal complexes and solid compounds *via* microbial cell-wall biomacromolecules, sometimes with further involvement of metal ions in cellular metabolism or their inactivation *via* special intracellular mechanisms, etc. (Langley and Beveridge 1999).

The bioavailability of iron, one of the few microelements vitally important for virtually all organisms (Sigel and Sigel 1998), which is commonly abundant in most soils, is yet well known to be extremely low within a wide range of environmental conditions. This is due to the very low solubility of ferric oxide or hydrated oxide species that are most stable under aerobic and microaerobic conditions. Thus, iron (III) largely represented by oxides or oxy hydroxides of variable crystallinity is an essential component of the majority of soil minerals. (iii) shuttle-assisted reduction involving specific compounds capable of redox cycling that serve as an electron shuttle between the bacterial cells and iron (III)-bearing mineral. All three mechanisms are clearly fundamentally different. Nevertheless, as for common features and distinctions, the first two mechanisms involve iron (III) reduction at the bacterial cell surface while the third one does not, which represents its main difference in the mode of electron transfer to iron(III). In addition to biotic (enzymatic) dissimilatory reduction by bacteria-reducers, humic substances, as well as their lower-molecular-weight precursors, are able to abiotically (chemically) reduce e.g., highly soluble Cr^{VI} It is interesting to note that some dissimilatory metal-reducing bacteria were found to be capable of sensing the depletion of soluble electron acceptors, with the subsequent synthesis of pili and flagella when growing on insoluble iron(III) or Mn(IV) minerals (but not on their soluble forms) (Lloyd 2003). Thereby the bacteria can adapt to the conditions by "switching on" their active chemotaxis mechanism to move towards the insoluble sources of electron acceptors (solid Fe^{III}- or Mn^{IV}-containing mineral particles), with subsequent attachment, thus passing from shuttle-assisted reduction (mechanism in view of shuttle deficiency, to contact reduction by mechanism).

The bacterial mobilization of heavy metals and radionucleoides are poorly understood processes Microorganisms may, in principle, be advantageous and biotechnologically beneficial when occurring in the rhizosphere of metal-tolerant plants (special hyperaccumulator plants), thereby facilitating phytoremediation processes. Interaction of metal ions with biological matter is essential as well as important for various biological processes for all organisms and in related fields (biogeochemistry, bioremediation and phytoremediation, biomining, biotechnology of metal extraction, sorption and recovery, etc). It can hardly be over-emphasized that a lack of understanding of molecular mechanisms underlying the effects of soil microorganisms and plant-root exudates on the state of metal compounds in the rhizosphere is a serious impediment in use of phyto-bioremediation technology" for cleaning soils contaminated with heavy metals. It can be overcome by widely using a variety of modern powerful physicochemical techniques in environmental and life sciences. On the other hand, knowing the mechanisms and routes of metal transformations may open ways for a variety of practical applications. It would be most imperative to genetically and/ or biotechnologically tailor plants for being hyper-accumulators, microbes to possess greater potential to convert toxic forms of heavy metals to less or non-toxic types and to harness the synergy of both partners to achieve better efficiencies for phyto-bioremediation of polluted soils through favorable plant-microbe interactions.

Chapter 2

MOLECULAR BIOLOGY

12. CHROMOSOME STRUCTURE: BASIC CHEMICAL ASPECTS – DNA, HISTONES AND NON-HISTONES; BASIC STRUCTURAL ASPECTS – THE NUCLEOSOMES, EUCHROMATIN AND HETEROCHROMATIN

Chromosomes

It contains the genetic material such as DNA and RNA. A chromosome is an organized package of DNA found in the nucleus of the cell. Different organisms have different numbers of chromosomes. Humans have 23 pairs of chromosomes--22 pairs of numbered chromosomes, called autosomes, and one pair of sex chromosomes, X and Y. Each parent contributes one chromosome to each pair so that offspring gets half of their chromosomes from their mother and half from their father (Reference: National Human Genome Research Institute).

Chromatin

The chromosomal material in its decondensed and thread-like state. Chromatin exists in two forms. One form, called euchromatin, is less condensed and can be transcribed. The second form, called heterochromatin, is highly condensed and is typically not transcribed. Under the microscope in its extended form, chromatin looks like beads on a string. The beads are called nucleosomes. Each nucleosome is composed of DNA wrapped around eight proteins called histones. The nucleosomes are then wrapped into a 30 nm spiral called a solenoid, where additional histone proteins support the chromatin structure. During cell division, the structure of the chromatin and chromosomes are visible under a light microscope, and they change in shape as the DNA is duplicated and separated into two cells (References: National Human Genome Research Institute and Scietable by nature education).

Mitosis

It forms asexual reproduction. It occurs when an organism grows or replaces damaged cells. Earlier to mitosis, a cell undergoes replication. The process in which chromatin is copied. It produces diploid cells. Other definition: A process of cell duplication, or reproduction, during which one cell gives rise to two genetically identical daughter cells. Strictly applied, the term mitosis is used to describe the duplication and distribution of chromosomes, the structures that carry the genetic information (https://www.britannica.com/science/mitosis) fig.12.1.

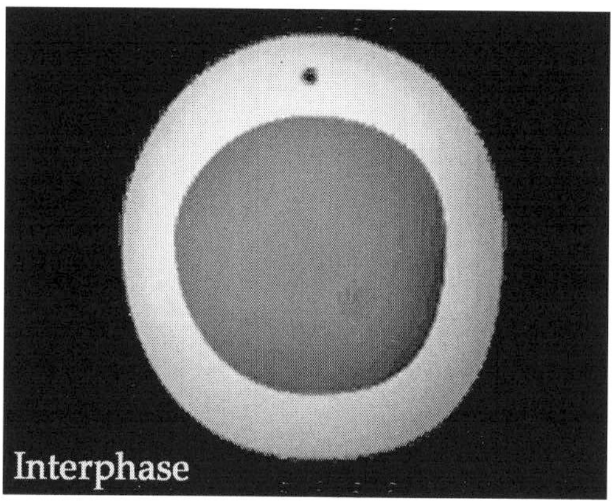

Figure 12.1. Interphase.

Prophase

Start of mitosis. The chromatin condenses into rod-like chromosomes. Each chromosome is consisting of sister chromatids, connected at the centromere and the nuclear membrane disappears fig.12.2.

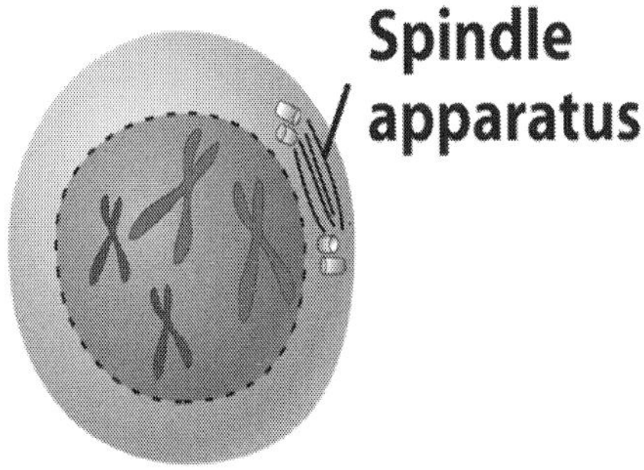

Figure 12.2. Prophase.

Metaphase

Chromosomes align themselves at the flat plane at the cell equator in metaphase fig.12.3.

Figure 12.3. Metaphase.

Anaphase

Centromeres are split. Sister chromatids-now chromosomes- are pulled to the opposite poles of the cell fig 12.4.

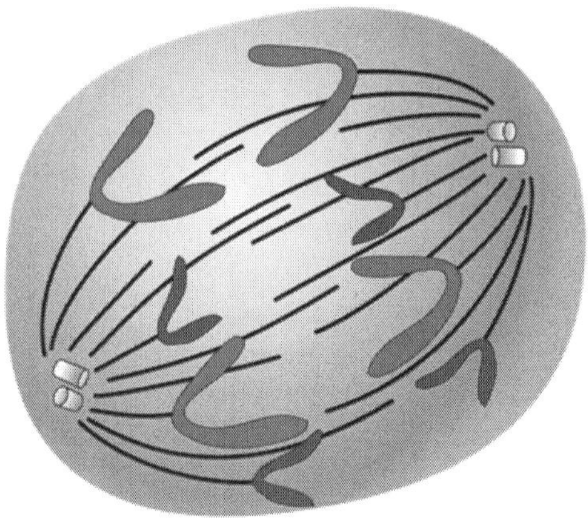

Figure 12.4. Anaphase.

Telophase

Chromosomes are unraveled, returning the chromatin to its non-dividing thread-like state. The nuclear membrane is assembled fig. 12.5

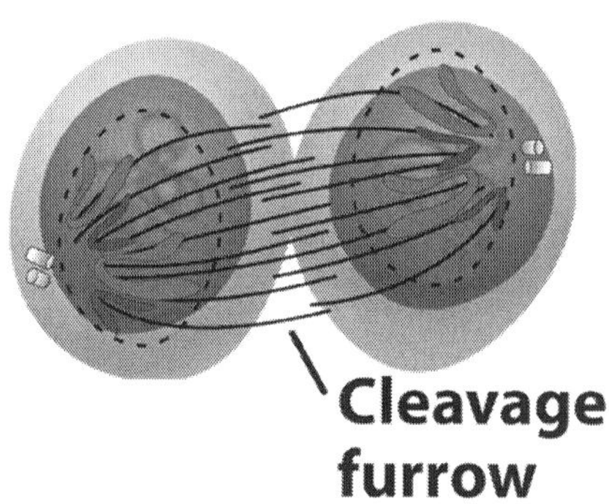

Figure 12.5. Telophase.

Cytokinesis

Division of the cytoplasm in cytokinesis and it initiates during anaphase and telophase fig. 12.6.

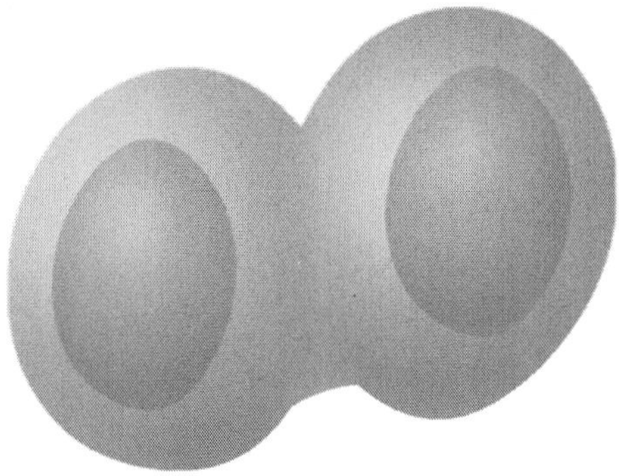

Figure 12.6. Cytokinesis.

Differs in animals and plant cells. Plant cells form a cell plate. Membranous vesicles congregate at the center of the cell. Vesicles are containing cell wall material. Animal cells form a cleavage furrow. It forms around the periphery of the dividing cell. It furrows become deeper and deeper until the membrane pinches off forming two cells. **Homologous pairs:** chromosomes that are closely matched in size and shape. It determines the same traits. **Sex chromosomes:** Those that determine the gender of the organism.

Chromosomal Theory of Inheritance

The two members of each pair of a homologous pair of chromosomes carry alleles for the same genes and, therefore, affect the same traits. It was proposed in 1903.

Mapping Genes

The location of a gene on a chromosome. It is currently trying to map all human genes to the appropriate chromosome. **Example:** the gene responsible for sickle cell anemia is located on chromosome 11.

Sexual Reproduction

The type of reproduction in which genetic information from females combines with males. It requires fertilization. The fusion of gametes (egg and sperm). These cells need to have half the amount of genetic information.

Meiosis

Meiosis, "on the other hand, is used for just one purpose in the human body: the production of gametes sex cells, or sperm and eggs. Its goal is to make daughter cells with exactly half as many chromosomes as the starting cell. Meiosis in humans is a division process that takes us from a diploid cell one with two sets of chromosomes to haploid cells ones with a single set of chromosomes. In humans, the haploid cells made in meiosis are sperm and eggs. When a sperm and an egg join in fertilization, the two haploid sets of chromosomes form a complete diploid set: a new genome (Khan Academy)». It produces haploid cells (gametes-eggs and sperm). It posses only one member of each pair of homologous chromosomes. The chromosomes replicate before meiosis. It requires two rounds of division.

Prophase I

The chromatin is condensing into compact chromosomes. The nuclear envelope disappears. The synapsis occurs. The homologous pairs of chromosomes closely align allowing the exchange of chromosome segments

Metaphase I

The aligned pairs of replicated chromosomes are move to the equator of the dividing cell fig.12.7.

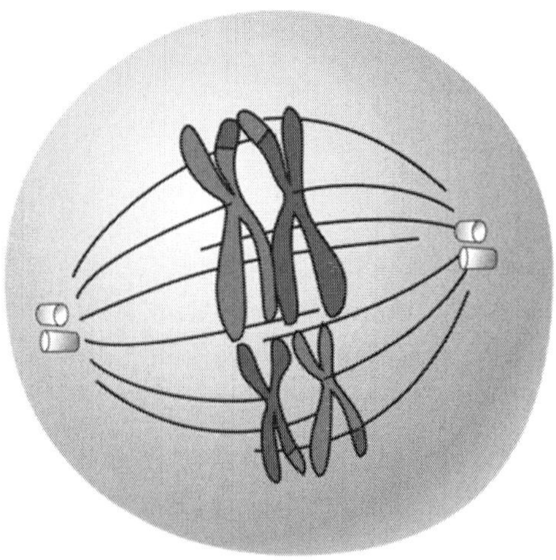

Figure 12.7: Metaphase.

Anaphase I

The members of homologous pairs of chromosomes are separate from each other. They move to the opposite poles of the cell. The chromosomes experience independent assortment fig 12.8.

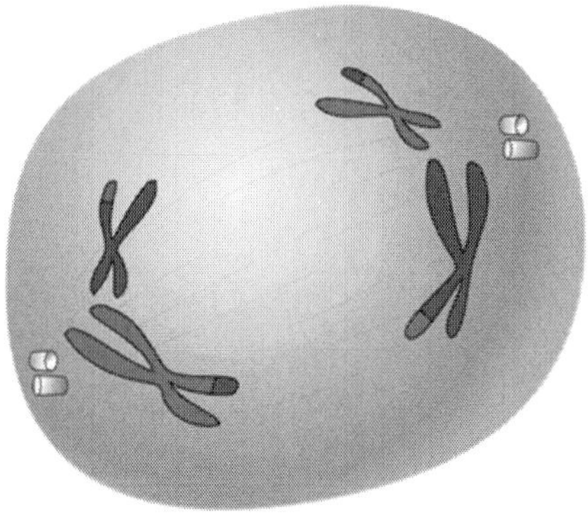

Figure 12.8. Anaphase I.

Telophase I

The chromosomes are cluster at opposite poles of the cell and begin to decondense. The nuclear envelope may reform. The cytokinesis occurs Interkinesis: phase between meiosis I and II. The daughter cells are now haploid fig. 12.9.

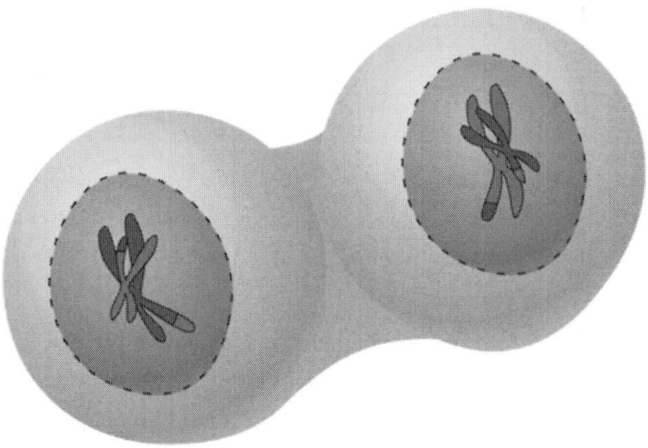

Figure 12.9. Telophase I.

Prophase Ii

The partially unraveled chromosomes condense again fig 12.10.

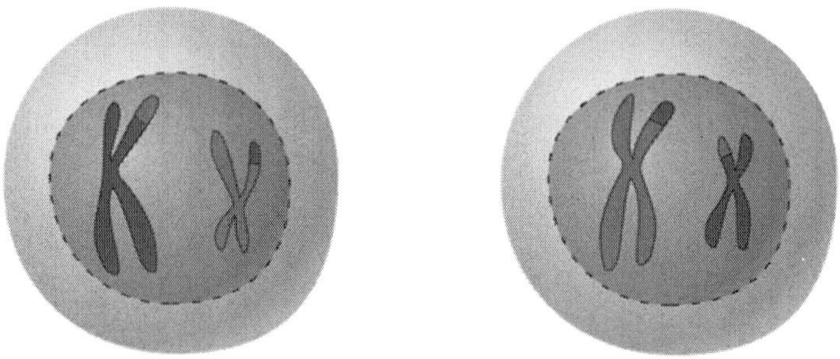

Figure 12.10. Prophase II.

Metaphase II

The chromosomes move to the cell equator. It has no longer homologous pairs, so chromosomes line up singly in the middle of the cell fig. 12.11.

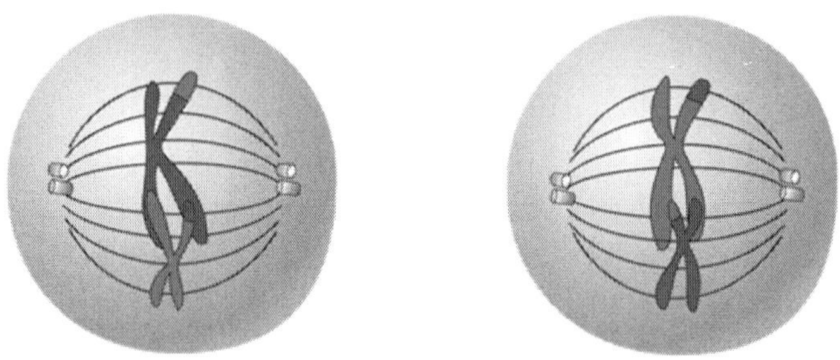

Figure 12.11. Metaphase II.

Anaphase II

The centromeres divide and chromosomes separate. It moves to the opposite poles of the cell fig. 12.12.

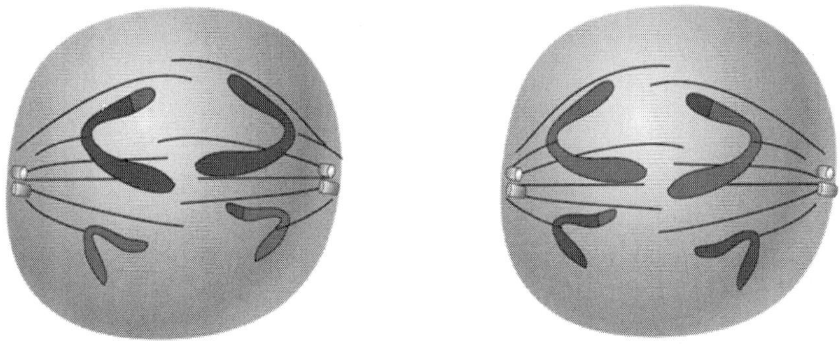

Figure 12.12. Anaphase II.

Telophase II

The clustered chromosomes at the cell pole begin to decondense. The nuclear membrane develops and cytokinesis occurs. It produces four non-identical haploid cells fig. 12.13.

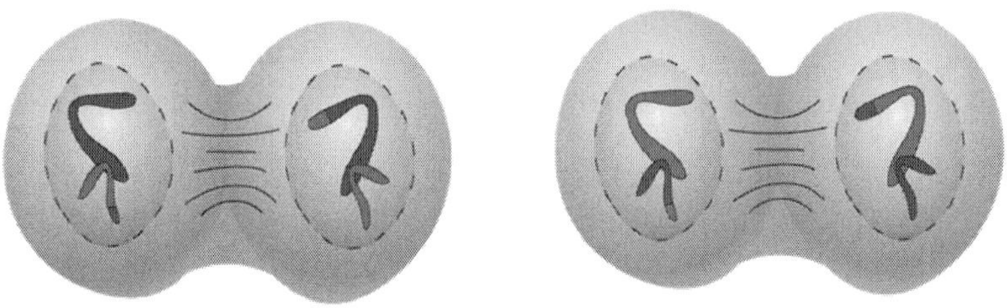

Figure 12.13. Telophase II.

Eukaryotic Chromosome Structure

Chromatin: It is a complex of DNA and chomosomal proteins and it is twice as much protein as DNA.

Two Major Types of Proteins

Histones: They are abundant and basic proteins with a positive charge that bind to DNA 5 main types: H1, H2A, H2B, H3, H4, equal in mass to DNA, evolutionarily conserved.

Non-histones: All the other proteins associated with DNA differ markedly in type and structure amounts vary widely, >> 100% DNA mass and **<< 50% DNA mass.**

Histone (*Components of the nucleosome*)

The major protein component of chromatin

↳ *Small, very basic (lys, arg) protein (to neutralize the highly charged DNA)*

↳ *assembly of histones with DNA requires, a molecular chaperone, nucleoplasmin*

Five Types of Histones:

H1 *Clamp for DNA wrapped around nucleosome*

"Core" Histones H2A, H2B, H3, H4 — *Found in 2 copies of each in the nuclosome. The DNA is then wrapped around the protein core mode of these proteins - together they form an octanner*

H4 *Is very highly concerned between species* (98% Identical between cones and peas) ~ 1% change in 600 million years

H3 Is also very conserved (97% identical)

Nucleasomal packaging of DNA

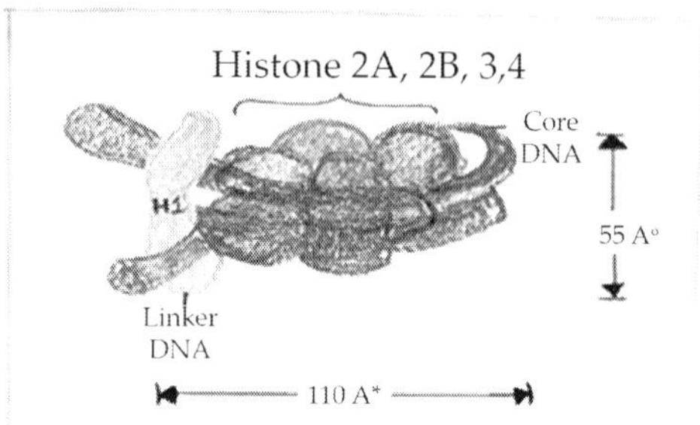

In the presence of Histone H1 175-200 bp DNA is associated with the nuclosome-but, only 1-16 bp is wrapped around the octomer (i.e-if H1 is removed ~ 146 bp) is observed in nuclease digits)

Histone H1 - helps 'champ' the DNA onto the nuclosome and participates in higher-order chromatin folding.

Histone modifications

Histones are also modified in order to regulate

1. chromosome strucluse; compaction
2. gene activity

The main types of modifications are

a) Phosphorylation of serines

b) Methylation of lysines

c) Acetylation of lysines : neutralies + Charges

$$-- NH_3^+ \rightarrow --NH-\underset{\underset{O}{\|}}{C}-CH_3$$

11nm fiber is coiled into 30 nm
Solenoid-requires histone H1

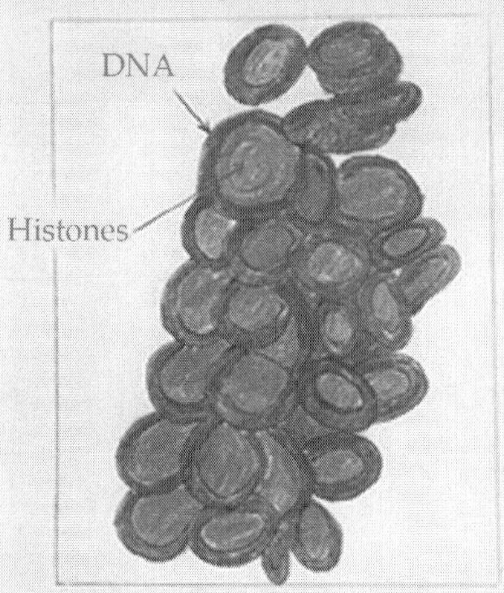

Schematic of formation
of 30 nm Chromatin fiber

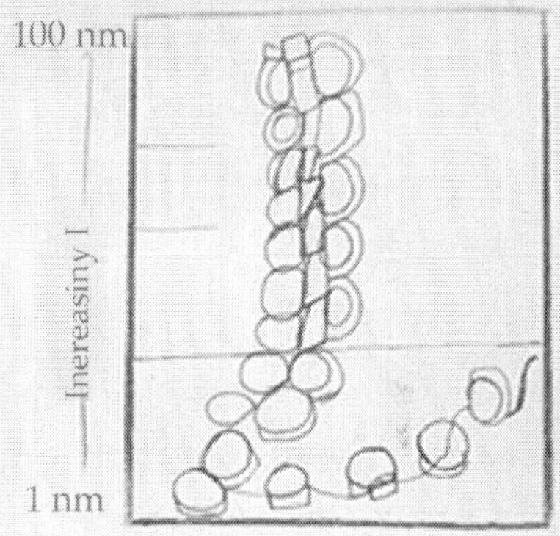

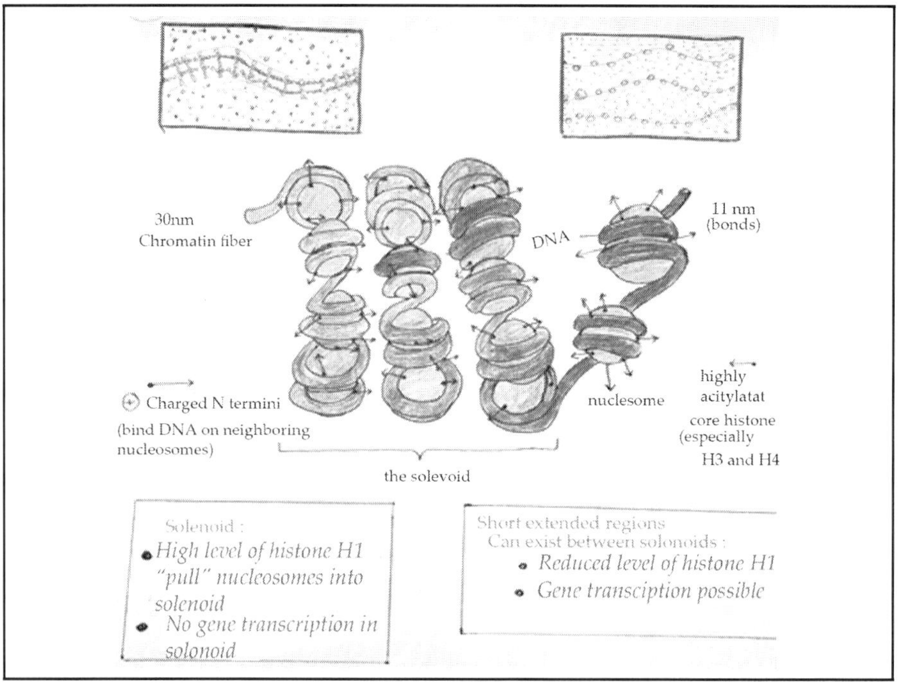
30nm
Chromatin fiber
DNA
11 nm
(bonds)
Charged N termini
(bind DNA on neighboring
nucleosomes)
nuclesome
highly
acitylatat
core histone
(especially
H3 and H4
the solevoid
Solenoid :
High level of histone H1
"pull" nucleosomes into
solenoid
No gene transcription in
solonoid
Short extended regions
Can exist between solonoids :
Reduced level of histone H1
Gene transciption possible

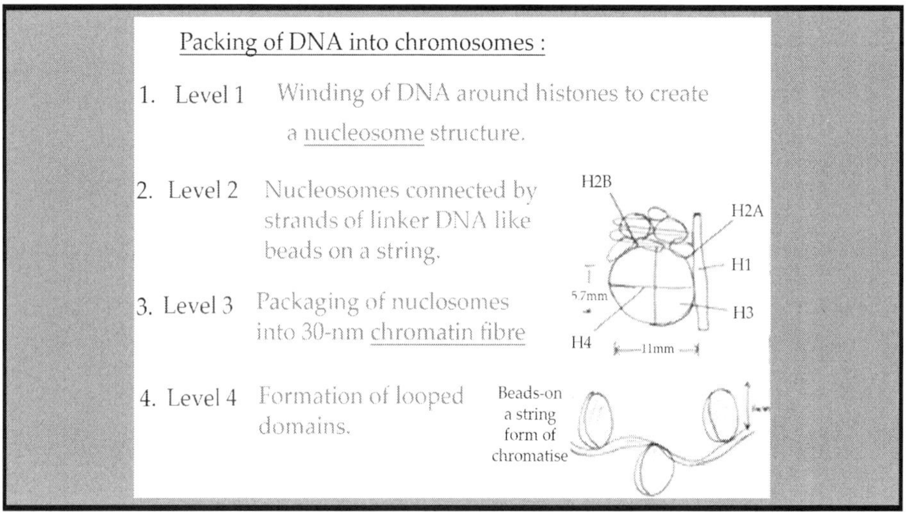
Packing of DNA into chromosomes :
1. Level 1 Winding of DNA around histones to create
a nucleosome structure.
2. Level 2 Nucleosomes connected by
strands of linker DNA like
beads on a string.
3. Level 3 Packaging of nuclosomes
into 30-nm chromatin fibre
4. Level 4 Formation of looped
domains.
H2B
H2A
H1
5.7mm
H3
H4
11mm
Beads-on
a string
form of
chromatise

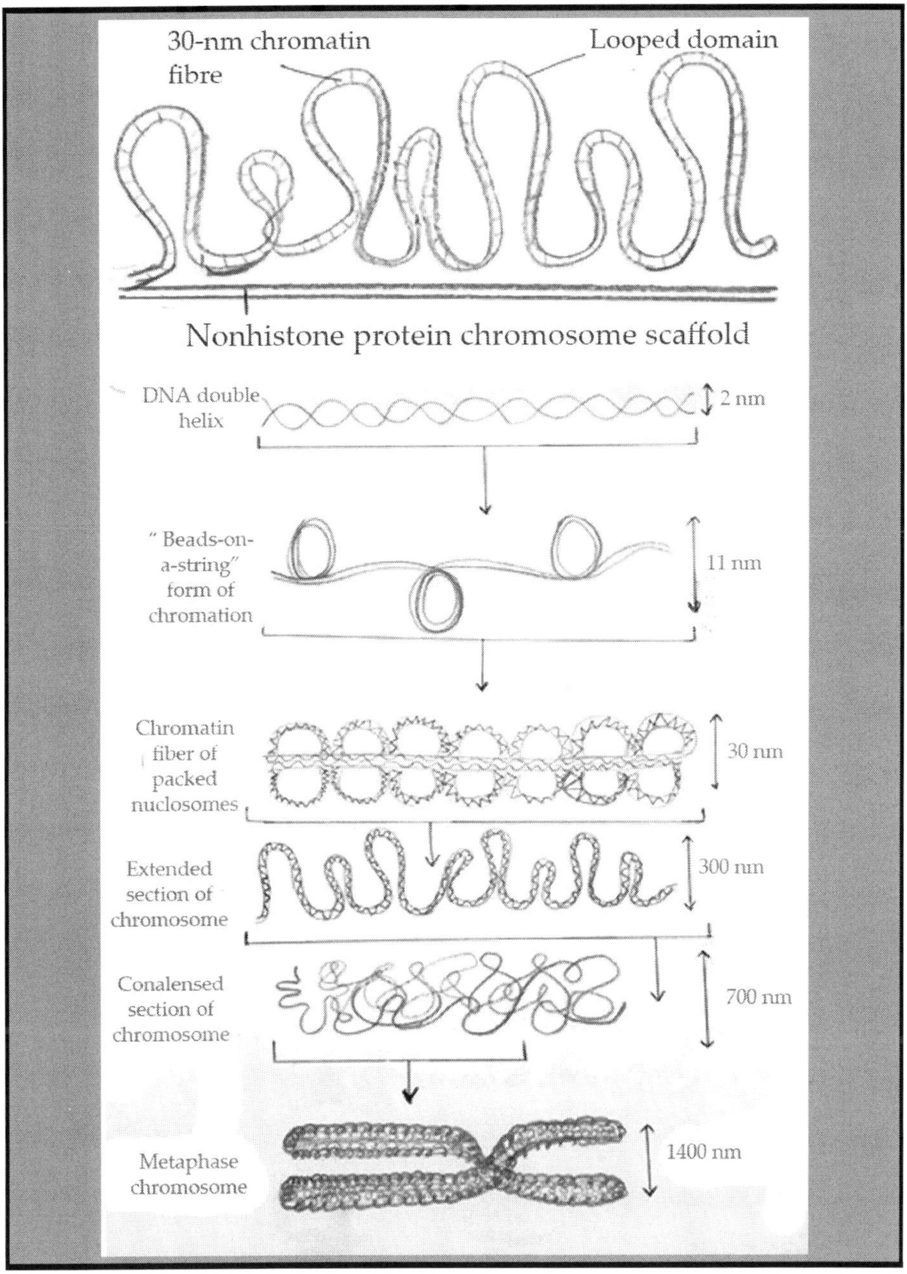

Chromatin Is Divided Into Euchromatin and Heterochromatin

Individual chromosomes can be seen only during mitosis. During interphase, the general mass of chromatin is in the form of euchromatin. Euchromatin is less tightly packed than mitotic chromosomes. The regions of heterochromatin remain densely packed throughout interphase.

Banding Patterns

Certain staining techniques cause the chromosomes to have the appearance of a series of striations called G-bands. The bands are lower in G • C content than the interbands. Genes are concentrated in the G • C-rich interbands.

Eukaryotic DNA Has Loops and Domains Attached To a Scaffold

DNA of interphase chromatin is negatively supercoiled into independent domains of ~85 kb. Metaphase chromosomes have a protein scaffold to which the loops of supercoiled DNA are attached. DNA is attached to the nuclear matrix at specific sequences called MARs or SARs. The MARs are A • T-rich but do not have any specific consensus sequence. The centromere is essential for segregation. A eukaryotic chromosome is held on the mitotic spindle by the attachment of microtubules to the kinetochore that forms in its centromeric region. The centromeres often have heterochromatin that is rich in satellite DNA sequences.

Telomeres

It seals the chromosome ends. The protein TRF2 catalyzes a reaction in which the 3 repeating unit of the G+T-rich strand forms a loop by displacing its homolog in an upstream region of the telomere.

Lampbrush Chromosomes

Sites of gene expression on lampbrush chromosomes show loops that are extended from the chromosomal axis.

Polytene Chromosomes

Polytene chromosomes of dipterans have a series of bands that can be used as a cytological map.

Bands that are sites of gene expression on polytene chromosomes expand to give "puffs." The nucleosome is the subunit of all chromatin. Micrococcal nuclease releases individual nucleosomes from chromatin as 11S particles. A nucleosome contains:

- ~200 bp of DNA

- It has two copies of each core histone (H2A, H2B, H3, and H4).
- One copy of H1.
- DNA is wrapped around the outside surface of the protein octamer.

DNA is coiled in arrays of nucleosomes which is greater than 95% of the DNA is recovered in nucleosomes or multimers when micrococcal nuclease cleaves DNA of chromatin. The length of DNA per nucleosome varies for individual tissues in a range from 154-260 bp.

Nucleosomes

Nucleosomal DNA is divided into the core DNA and linker DNA depending on its susceptibility to micrococcal nuclease. The core DNA is the length of 146 bp that is found on the core particles produced by prolonged digestion with micrococcal nuclease. Linker DNA is the region of 8-114 bp that is susceptible to early cleavage by the enzyme. It changes in the length of linker DNA account for the variation in the total length of nucleosomal DNA. H1 is associated with linker DNA and may lie at the point where DNA enters and leaves the nucleosome.

Polytene Chromosome and Lambrush Chromosome

Giant chromosomes were first observed by E. G. Balbiani in 1881 in the salivary gland cells of Dipteran species. Polytene chromosomes are giant chromosomes found in the salivary glands, malpighian tubules, gut lining and adipose tissue cells of some Dipteran larvae. These chromosomes may reach a size up to 200 times (or more) the size of corresponding chromosomes at meiosis or in nuclei of normal mitotic cells. These giant chromosomes are formed by repeated cycles of endo-reduplication of single chromatids. This means that the chromatids replicate without cell division, as a result, the number of chromatids keeps increasing. This process is called polyteny and therefore, these chromosomes are called polytene chromosomes.

Polytene chromosomes show another characteristic feature i.e., somatic pairing. As a result of this, chromosomes always appear to be half of that in normal somatic cells. Lampbrush chromosomes were first seen in salamander oocytes by Fleming in 1882. The name lampbrush was given by Ruckert in 1892. The nomenclature comes from its similarity in appearance to the brushes used to clean lamp chimneys in old days. These chromosomes are transitory structures that exist during the extended Diplotene in oocytes of most invertebrates and vertebrates except mammals. The loops in these chromosomes are sites of RNA transcriptions. The condensed chromatin in the chromomeres is generally not transcribed. The position of each loop is fixed. A given loop always contains the same DNA sequence. They have been widely used in studies of chromosome organization, gene expression, and molecular morphology of RNA transcription.

13. DNA REPLICATION: DNA REPLICATION AND ITS OVERVIEW: SEMICONSERVATIVE MECHANISM OF DNA REPLICATION, REPLICONS, ORIGINS AND TERMINI, SEMI-DISCONTINUOUS REPLICATION, RNA PRIMING, REPLICATION IN PROKARYOTES AND EUKARYOTES

Introduction

The DNA is found in all plants, animals, prokaryotes and some viruses. DNA is a double-helical structure that consists of genetic material. It is joined by a phosphodiester bond. It is acidic in nature. The entire genetic message that controls the chemistry of every cell of the body acting in a specific way is actually written in the language of four nitrogenous bases of DNA i.e., purines and pyrimidines.

Composition of DNA

It is a polymeric compound containing four monomers known as deoxyribonucleotide monomers or deoxyribotides. Each DNA consists of a pentose sugar, a phosphate group and nitrogenous bases: **Figure 13.1.** Purines bases (adenine and guanine) are heterocyclic and two ringed bases and the pyrimidines (thymine and cytosine) are one ringed bases.

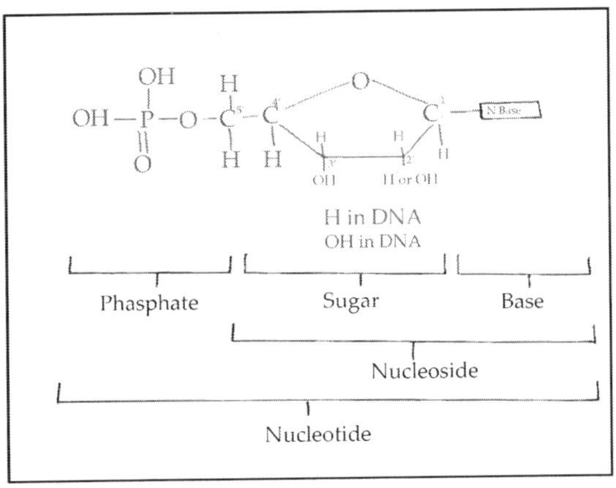

Figure 13.1. Nucleotides.

Nucleosides

The nitrogenous bases are combined with a pentose sugar are called nucleosides.

Nucleotides

A nucleoside linked with phosphate forms a nucleotide. Nucleoside has pentose sugar + nitrogenous base nucleoside + phosphate Nucleotide. On the basis of different nitrogenous bases, the deoxynucleotides are of following types: (i) Adenine (A) deoxyadenosine-3'/5'-monophosphate (3'75'-d AMP) deoxyguanosine -5'-monophosphate (5'-d GMP) deoxythymidine -5-monophosphate (5'-d TMP) (ii) Guanine (G) (iii) Thymine (T) (iv) Cytosine (C) deoxyeytidine -5-monophosphate (5'-d CMP). In addition to the presence of nucleosides in the DNA helix, these are also present in nucleoplasm and cytoplasm in the form of deoxyribonucleotide phosphates eg. deoxyadenosine triphosphate (dATP), deoxyguanosine triphosphate (dGTP), deoxycytidine triphosphate (dCTP), deoxythymidine triphosphate (dTTP) **figure 13.2**. The advantage of these four deoxyribonucleotides in triphosphate forms is that the DNA polymerase acts only on triphosphates of nucleotides during DNA replication. Similarly, the ribonucleotides contain ribose sugar, nitrogenous bases and phosphate. Except for sugar, the other components are similar. However, uracil (U) is found in RNA instead of thymine. Generally, an RNA molecule is single-stranded besides some exceptions.

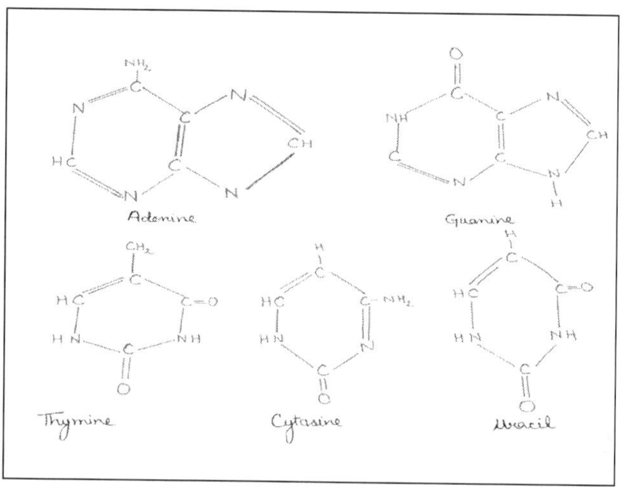

Figure 13.2. Nitrogenous base.

Polynucleotide

The nucleotides undergo the process of polymerization to form a long chain of polynucleotides. The nucleotides are designated by prefixing 'poly' to each repeating unit such as poly A (polyadenylic acid), poly T (polythymidilic acid), poly G (polyguanidylic acid), poly C (polycytidilic acid) and poly U (poly uridylic acid). The polynucleotides that consist of the same are called homopolynucleotides such as poly A, poly T, poly G, poly C and poly U.

Chargaff-Equivalence Rule

By 1948, chemist **Erwin Chargaff** started using paper chromatography to analyze the base composition of DNA from a number of studies. In 1950, **Chargaff** discovered that in the DNA of different types of organisms: the total amount of purines is equal to the total amount of pyrimidines i.e. The total number of A is equal to the total number of T (A-T), and the total number of G is equal to the total number of C (G-C). It means that A/T = G/C i.e. A+T/G+C 1. In the DNA molecules isolated from several organisms, regularity exists in the base composition.

Physical Nature of DNA

Watson and Crick's Model of DNA

J.D. Watson and F.H.C. Crick (1953) combined the physical and chemical data generated by earlier workers and proposed a double-helix model for the DNA molecule. This model is widely accepted. According to this model, the DNA molecule consists of two strands that are connected together by hydrogen bonds and helically twisted. Each step on one strand consists of a nucleotide of purine base that alternates with that of a pyrimidine base. Thus, a strand of DNA molecule is a polymer of four nucleotides i.e. A. G. T, C. The two strands join together to form a double helix. Bases of two nucleotides form hydrogen bonds i.e A combines with T by two hydrogen bonds (A T) and G combines with C by three hydrogen bonds (G C). However, the sequence of bonding is such that for every A.T.G.C. on one strand there would be T, A, C, G on the other strand. Therefore, the two chains are nucleotides on one chain are the photocopy of the sequence of nucleotides on the other chain.

The two strands of double helix run in antiparallel direction i.e. they have opposite polarity. The left-hand strand has 5'-3' polarity, whereas the right hand has 3’-5’ polarity to the first one. The polarity is due to the direction of phosphodiester linkage complementary to each other i.e. sequence of as compared to the first one. The DNA model also suggested a copying mechanism of the genetic material. DNA replication is the fundamental and unique event underlying growth and reproduction in all living organisms ranging from the smallest viruses to the most complex of all creatures including man. DNA replicates by a semiconservative mechanism that was experimentally proved by **Mathew, Meselson and Frank W. Stahlin** in 1958. If changes occur in the sequence or composition of base pairs of DNA, mutation takes place. Though the presence of adenine, guanine, thymine and cytosine is a universal phenomenon, yet unusual bases in DNA molecule also occur. In some bacteriophages, 5- hydroxymethyl- cytosine (HMC) replaces cytosine of the DNA molecule when methylation of adenine, guanine and cytosine occurs. This results in changes in these bases: **figure 13.3.**

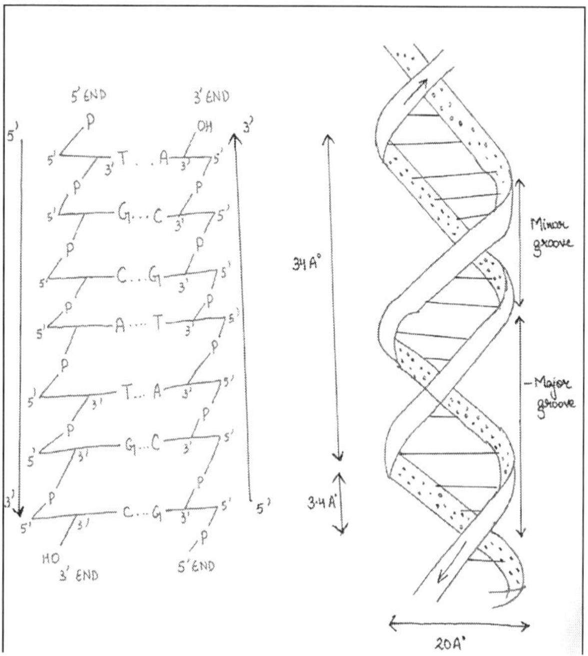

Figure 13.3. Watson crick’s model of DNA.

Three Major Forms

A-DNA

B-DNA

Z-DNA

B-DNA is biologically the most common

- Right-handed (20 angstrom (Ao) diameter)
- Complementary base-pairing (Watson-crick)
- – A-T = G-C
- Each base pair has the same width
  10.85 Ao from c1' to c1'
  A-T and G-C pairs are interchangeable.

"Pseudo-dyad" axis of symmetry

Ideal **B-DNA** has 10 base pairs per turn and base thickness. The aromatic rings with 3.4 A° **thickness to rings.**

- Pitch = 10 x 3.4 = 34 A° per complete turn.
- Axis passes through the middle of each bp
- Minor groove is narrow.
- Major groove is wide
- Class exercise: explore the structure of B-DNA. Pay special attention to the major, minor grooves.
- Minor: exposes edge from which C1' atoms extend.
- Major: exposes the opposite edge of the base pair.
- The pattern of H-bond possibilities is more specific and more discriminating in the major groove.

A-DNA

It is a right-handed helix. It is wider and flatter than B-DNA 11.6 bp per turn. The pitch of 34 A°. An axial hole. The base planes are tilted 20 degrees with respect to the helical axis. The helix axis passes "above" the major groove. The deep major and shallow minor groove. It was observed under dehydrating conditions. When relative humidity is ~ 75%. B-DNA → A-DNA (reversible). Most self-complementary oligonucleotides of < 10 bp crystallize in A-DNA conf. A-DNA has been observed in 2 contexts:

- At active site of DNA polymerase (~ 3 bp)
- Gram (+) bacteria undergoing sporulation

It induces B-DNA to → A-DNA: resistant to UV-induced damage and cross-linking of pyrimidine bases.

Z-DNA

A left-handed helix. It is seen in conditions of high salt concentrations. It reduces repulsions between closest phosphate groups on opposite strands (B-DNA) in complementary polynucleotides with alternating purines and pyrimidines.

- poly d(GC) · poly d(GC)
- poly d(AC) · poly d(GT)

It might also be seen in DNA segments with the above characteristics with 12 base pairs per turn. A pitch of 44 degrees. A deep minor groove. No discernible major groove. The reversible change from B-DNA to Z-DNA in localized regions may act as a "switch" to regulate gene expression. Transient formation behind actively transcribing RNA polymerase.

Palindromes in DNA Sequences

The genetic palindromes are similar to verbal palindromes. A palindromic sequence in DNA is one in which the 5' to 3' base pair sequence is identical on both strands (the 5' and 3' ends refer to the chemical structure of the DNA).

RNA is single-stranded and does not have to stay in the nucleus. RNA is not found in chromosomes because it does not carry the genetic code; however, it can read the DNA code and take the information out of the nucleus.

- RNA's main work is to build proteins.
- The building blocks of RNA are nucleotides, just like DNA.

A Nucleotide in RNA is still made of 3 important things:

1. 5-Carbon Sugar - **<u>Ribose</u>** (instead of Deoxyribose)
2. Phosphate
3. Nitrogen base

There are 4 nitrogen bases in RNA, A, G, C, and <u>U</u> that pair together)

A →**<u>U</u>** C →G

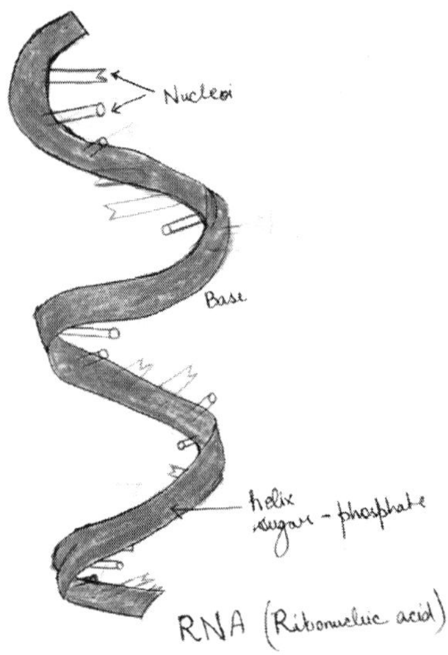

Three Main Types of RNA

Messenger RNA (mRNA): It carries copies of instructions for the assembly of amino acids into proteins from DNA to the rest of the cell (serve as a "messenger").

Ribosomal RNA (rRNA): It makes up the major part of ribosomes, which is where proteins are made.

Transfer RNA (tRNA): The transfers amino acids to ribosomes during protein synthesis.

DNA Denaturation

Strands are held together by weak, noncovalent bonds. The strands start to separate at a specific temperature.

- Within a few degrees, the process is complete.
- The solution contains single-stranded molecules.

- Higher single strand absorbance at 260 nm.
- Hydrophobic base interactions are reduced.
- Bases absorb photons more efficiently.

Melting temperature (T_m) is half denatured. T_m increases with G-C content (%G + %C). A-T-rich sections melt before G-C-rich segments.

DNA Renaturation

J. Marmur (Harvard, 1960) – first described slowly cool heat-denatured DNA. The temperature quickly is brought to ~25°C below T_m& incubate a while complementary single-stranded DNAs can reassociate or reanneal. Renaturation is very useful for genome complexity: variety & copy number and hybridization: molecular identification. Renaturation is governed by ionic strength of the solution, Temperature, Time, DNA concentration and size of the interacting molecules

Smaller Genome - Faster Renaturation

More copies of small genomes and increases the chance of collision between complementary fragments and viral/bacterial genomes are symmetrical curves.

Circular and Superhelical DNA

Almost in all the prokaryotes and a few viruses, the DNA is organized in the form of closed circles. The two ends of the double helix get covalently sealed to form a closed circle. Thus, a closed circle contains two unbroken complementary strands. Some times one or more nicks or breaks may be present on one or both strands, for example, DNA of phage PM, Besides some exceptions, the covalently closed circles are twisted into super helix or super coils and is associated with basic proteins but not with histones found complexed with all eukaryotic DNA these histones like proteins appears to help the organization of bacterial chromatin structure with the result of nucleosome like structure, folding and supercoiling of DNA and association of DNA polymerase with nucleoids. These nucleoid-associated proteins include HU proteins, IHF, proteins HI, Fir A, H-NS and Fis. In archaeobacteria (e.g. Archaea) the chromosomal DNA exists in protein associated form. Histone-like proteins have been isolated from nucleoprotein complexes in *Thermoplasma acidophilum* and *Halobacterium salinarum*.

Thus, the protein associated DNA and nucleosome like structures are present in a helix coils clockwise from the axis the coiling is termed as positive contrast, if the path of coiling is anticlockwise, the coil is called left-handed or negative coil. The two ends of a linear DNA helix can be joined to make each strand continuous. However, if one end rotates at 360° with respect to the other. These histones like proteins are appearing to help the organization of bacterial DNA into a coiled variety of bacteria. The unwinding of the double joined resulting in the formation of a twisted circle in opposite sense i.e. opposite helix; the ends are to unwinding direction.

Such a twisted circle appears as an 8 it has one node or crossing over point. If it is twisted at 720° before joining, the resulting super helix will contain two nodes. The enzyme topoisomerases alter the topological form i.e. super coiling of a circular DNA molecule. Type I topoisomerases (e.g. *E. coli* top A) relax the negatively super coiled DNA by breaking the 5'-phosphoryl end and then resealing the nicked phosphodiester backbone (Moat and Foster, 1995). Type II Topoisomerases need energy to unwind the DNA molecules resulting in the introduction of super coils. One type of type II isomerases, DNA gyrase is apparently responsible for the negatively super coiled state of the bacterial chromosome. Super coiling is essential for efficient replication and transcription of prokaryotic DNA. The bacterial chromosomes are believed to contain about 50 negatively super coiled loops topological unit, the boundaries of which may be defined by the sites on DNA that limit its rotation (Wang, 1982; Mout and Foster, 1995). One of the phosphodiester bonds in dsDNA allows the 3'-OH end to swivel around or domains.

Organization of DNA

In addition to the organization of DNA in prokaryotes and lower eukaryotes as discussed earlier in eukaryotes, the DNA helix is highly organized into the well-defined DNA-protein complex termed as nucleosomes. Among the proteins, the most prominent are histones. The histones are small and basic proteins rich in amino acids such as cells there are four types of histones e.g. H2A, H2B, H3, and H4. Eight histone molecules (two each of H2A, H2B, H3, and H4) form an octamer ellipsoidal structure of about 1 nm to 7 nm in diameter. DNA coils around the surface of the ellipsoidal structure of histones 166 base pairs (about 7/4 turns) before proceeding onto the next and form. Thus, a nucleosome is an octamer or four histone proteins complexed with DNA lysine and/or arginine. Almost in all eukaryotic long and 6.5-complex structure, the nucleosome.

The histones play important role in determining eukaryotic chromosomes by determining the conformation known as chromatin. The nucleosomes are the repeating units of DNA organization which are often termed as beads. The DNA isolated from chromatin looks like a string of 146 base pairs of DNA lie in the helical path and the histone. DNA assembly is known as the nucleosome core. The stretch of DNA between the nucleosome is known as linker' which varies in length from 14 to over 100 base pairs. The H_1 is associated with the linker region and helps the folding of DNA into a complex structure called chromatin or beads. As a result of m chromatin becomes visible as chromosomes during cell division.

Structure of Ribonucleic Acid (RNA)

The RNA TMV, yellow mosaic virus, influenza virus, foot and mouth disease is usually single-stranded except one virus such as virus, etc. which have dsDNA. The retrovirus is an entirely single strand of the RNA which folded either at certain regions or to form a hairpin-shaped structure. In the hairpin-shaped structures, the complementary bases are linked by hydrogen bonds which give stability to the molecules. However, no complementary bases are found in the unfolded region. The RNA does not possess an equal purine-pyrimidine ratio, as it is found in the DNA. Like DNA, the RNA is also the polymer of four nucleotides each one contains D-ribose, phosphoric acid and a nitrogenous base. The bases are two purines (A, G) and two are not found in RNA. A pairing between bases occurs as A-U and G-C pyrimidines (C, U). Thymine is not found in RNA. The pairing is between bases occurs as A-U and G-C. The nucleotides formed by the four bases are adenosine monophosphate (AMP), guanosine monophosphate (GMP), cytosine monophosphate (CMP), and uridine monophosphate (UMP). These are found freely in nucleoplasm but in the form of triphosphates e.g. ATP, GTP, UTP and CTP. As a result of polymerization, the ribonucleotides form a polynucleotide chain of RNA. If the RNA is involved in genetic mechanisms" it is called genetic RNA as found in plants, animal and bacterial viruses. The DNA acts as genetic material and RNA follows the order of DNA. In such cells, the RNA does not have a genetic role. Therefore, it is called non-genetic RNA. The non-genetic RNA is of three types: (i) ribosomal RNA (rRNA), (ii) transfer RNA (1RNA) or soluble RNA (sRNA), and (iii) messenger RNA (mRNA) or template RNA. These three types of RNA differ from each other in structure, site of synthesis in eukaryotic cell and function.

DNA Replication

The transmission of chromosomal DNA from generation to generation is crucial to cell propagation. This can only be achieved when chromosomal DNA is accurately replicated, providing two copies of the entire genome for faithful distribution into each daughter cell. Cells need to make a copy of DNA before dividing so each daughter cell has a complete copy of genetic information.

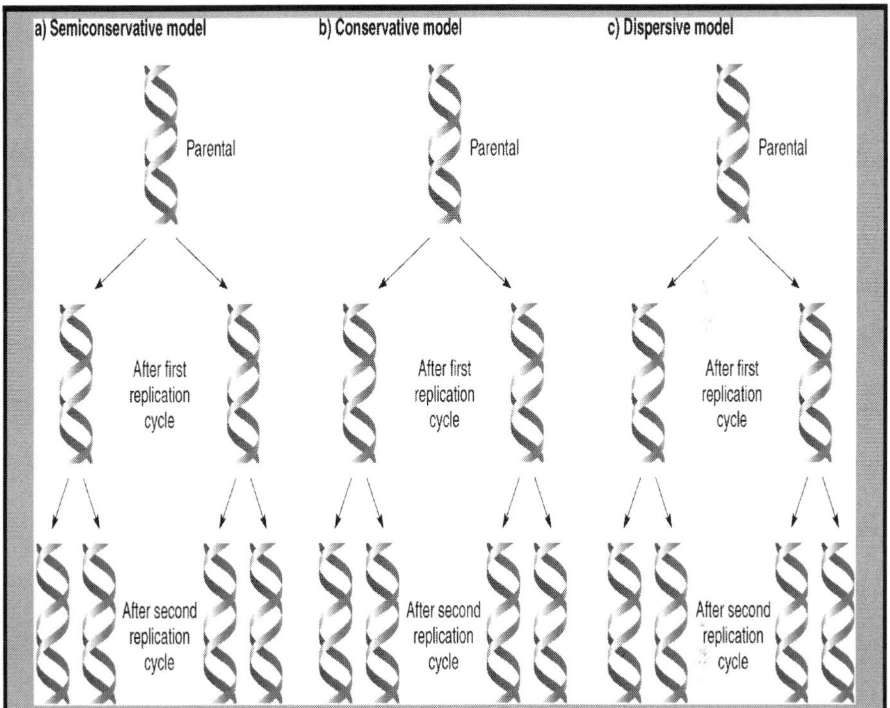

Semi-conservative Model

Replication of DNA

Base pairing allows each strand to serve as a template for a new strand. The new strand is 1/2 parent template & 1/2 new DNA. It has anti-parallel strands. The nucleotides in the DNA backbone are bonded from phosphate to sugar between 3′& 5′ carbons.

- DNA molecule has "direction"
- Complementary strand runs in the opposite direction

Meselson and Stahl's Experiment

Two isotopes of nitrogen:

^{14}N common form; ^{15}N rare heavy form.

- *E. coli* were grown in a ^{15}N media first, and then transferred to ^{14}N media.
- Cultured *E. coli* were subjected to equilibrium density gradient centrifugation.

Modes of Replication

Replicons: units of replication. Theta replication: circular DNA, *E. coli*; the single origin of replication forming a replication fork, usually a bidirectional replication. Rolling-circle replication: virus, F factor of *E. coli* and a single origin of replication.

Linear Eukaryotic Replication

Eukaryotic cells; thousands of origins; a typical replicon: 200,000 ~ 300,000 bp in length. Linear eukaryotic replication.

Direction of Replication

DNA polymerase adds nucleotides only to the 3′ end of a growing strand. The replication can only go 5′→3′.

Bonding in DNA

Hydrogen bonds and covalent phosphodiester bonds.

Replication: 1st Step

Initiation: 245 bp in the *oriC* (single origin replicon); an initiation protein. The unwinding of DNA is performed by ***Helicase***. ***Gyrase*** removes supercoiling ahead of the replication fork. Single-stranded DNA is prevented from annealing by single-stranded binding proteins. Primers: an existing group of RNA nucleotides with a 3′-OH group to which a new nucleotide can be added; usually 10 ~ 12 nucleotides long.

Primase: RNA polymerase

Unwind DNA

Helicase enzyme unwinds part of DNA helix. Stabilized by single-stranded binding proteins.

Prevents DNA molecule from closing

DNA gyrase is an enzyme that prevents tangling upstream from the replication fork.

Replication: 2nd Step

RNA Primase. Adds a small section of RNA (RNA primer) to the 3' end of template DNA. DNA polymerase 3 (the enzyme that builds a new DNA strand) can only add nucleotides to existing strands of DNA.

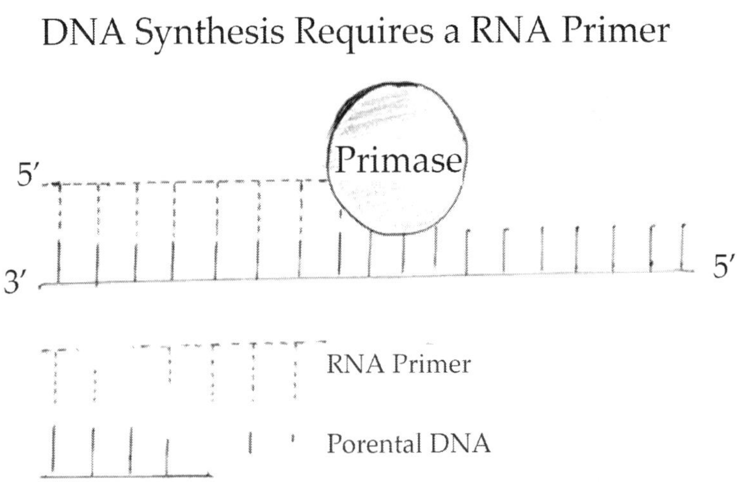

Elongation

It is carried out by DNA polymerase III. The removing RNA primer: DNA polymerase I. DNA ligase: connecting nicks after RNA primers are removed. **Termination:** when a replication fork meets or by a termination protein.

Replication: 3rd Step

Build the daughter DNA strand. Add new complementary bases. With the help of the enzyme DNA polymerase III.

Replication: 4th Step

Replacement of RNA primer by DNA and is done by DNA polymerase I.

Direction of Replication

Leading strand: It undergoes continuous replication. **Lagging strand:** It undergoes discontinuous replication. **Okazaki fragments:** The discontinuously synthesized short DNA fragments forming the lagging strand.

Leading & Lagging Strands

Limits of DNA polymerase III. It can only build onto 3′ end of an existing DNA strand.

Lagging Strand

Okazaki fragments. It is joined by ligase and "spot welder" enzyme.

Leading Strand

It is produced by continuous synthesis.

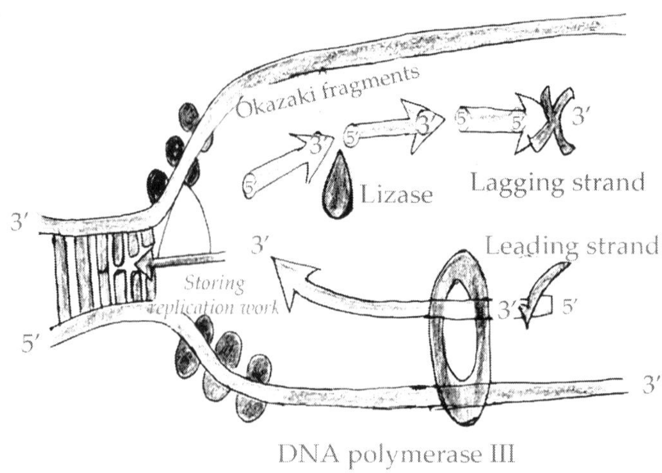

DNA Replication on the Lagging Strand

RNA primer is added. It is made by primase. It serves as a starter sequence for DNA polymerase III. However, short segments called Okazaki fragments are made because the enzyme can only go in a 5' →3' direction. Replacing RNA primers with DNA.

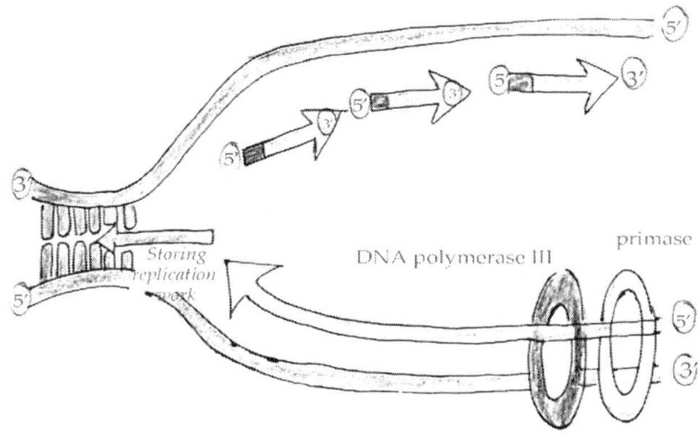

DNA Polymerase I

It removes sections of the RNA primer and replaces them with DNA nucleotides.

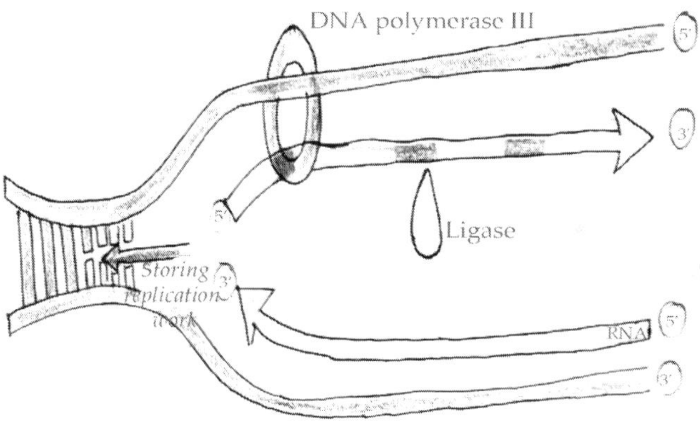

Replication Fork

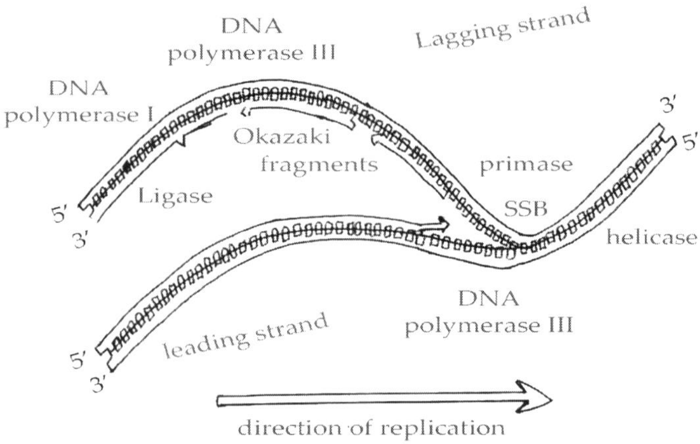

SSB = single-stranded binding proteins.

DNA polymerase delta and epsilon are attached to the DNA and held in place by other proteins. Sliding-clamp proteins (proliferating cell nuclear antigen - PCNA), allow the stable binding of DNA polymerase and strand synthesis. Clamp-loading proteins (replication factor C - RFC), aid in attaching the sliding-clamp proteins. The DNA is copied by a replication machine that travels along each replication fork. One of the most important members of this complex is DNA polymerase. It is able to add DNA subunits and make new DNA.

DNA Polymerase

It was discovered in 1956 by **Kornberg** first time from *E. coli* bacteria. DNA Pol I, II, and III. DNA Pol III is involved in replication of DNA. DNA Pol I involved in repair. Humans have 4 types of polymerase. DNA Pol alpha, beta, delta - nuclear DNA. DNA Pol gamma - mitochondrial DNA. All DNA Pols have 2 properties. Only synthesize DNA in one direction 5' to 3'. Only add to the end of existing double-stranded DNA. Therefore, they cannot start the synthesis of DNA from scratch. RNA polymerases can, but not DNA polymerases.

It has a proofreading mechanism built. Checks for base matching. Removes mismatched bases by going backward. The reason why it is not able to build DNA in the 3' to 5' direction. No energy from ATP hydrolysis. Makes just one error in 10E8 or 10E9 (billion) bases added. The unwinding of DNA strands by helicases. Require ATP. Single-stranded DNA-binding proteins. Bind to single-stranded DNA to stabilize the structure. RPA (replication protein A - in eukaryotes). Topoisomerase - helps with the prevention of DNA strand twisting - 'swivels' and two types.

Type I - Break one strand only and then rejoins.

Type II - Break both strands and then rejoin.

Telomerase

It is an enzyme with a role as a reverse transcriptase. It acts without a DNA template, as it has its own RNA template stored inside.

The Fidelity of DNA Replication

Proofreading: DNA polymerase I: $3' \rightarrow 5'$ exonuclease activity removes the incorrectly paired nucleotide.

Mismatch repair: correcting errors after replication is complete.

Eukaryotic DNA Replication

Eukaryotic DNA polymerase. DNA polymerase α- acts like Primase to initiate. DNA polymerase δ- replicates the lagging strand. DNA polymerase e- replicates the leading strand.

Replication at the Ends of Chromosomes

Telomeres and telomerase. DNA repair mechanism. Damaged DNA must be repaired. If the damage is passed on to subsequent generations, then we use the evolutionary term - mutation. It must take place in the germ cells - the gametes -eggs and sperm. If the damage is to somatic cells (all other cells of the body bar germ cells) then just that one individual is affected. Damage from where consequences of DNA replication errors. Chemical agents acting on the DNA.UV light imparting energy into a DNA molecule. The spontaneous changes to the DNA.

Repair Mechanisms

1. Reversal of damage
2. Excision repair
3. Mismatch repair
4. Recombination repair

5. Error-prone repair
6. Restriction-modification systems

Enzymatically Undo the Damage

a) Photoreactivation

b) Removal of methyl groups

Photolyase Breaks Apart Pyrimidine Dimers

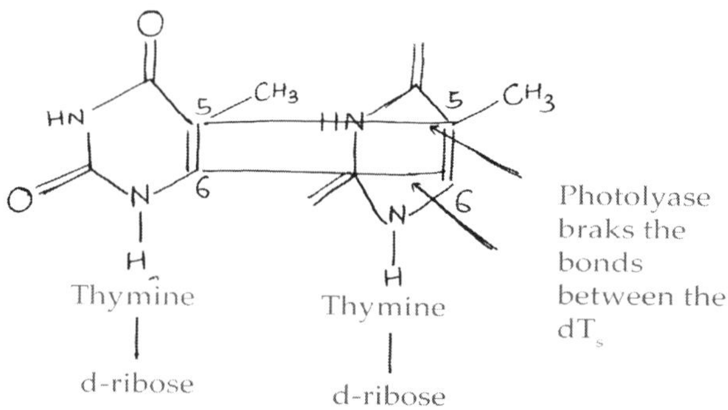

1A. Photoreactivation

Photolyase: It binds pyrimidine dimers and catalyzes a photochemical reaction. It breaks the cyclobutane ring and reforms two adjacent T'. There are 2 subunits and encoded by phrA and phrB.

Conversion Of 6-O-Methyl-Dg Back To Dg

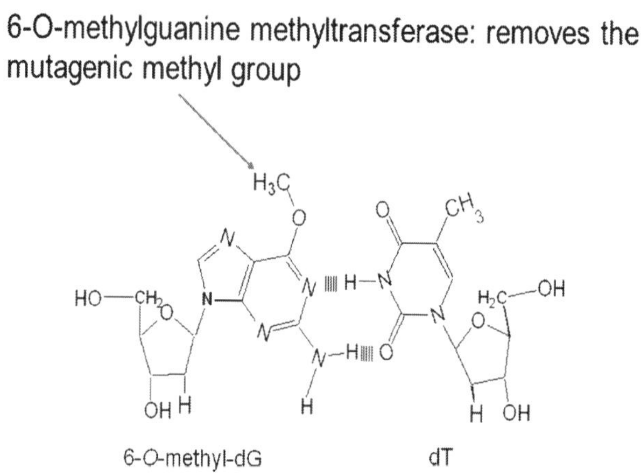

1B. Removal of Methyl Groups

6-O-methylguanine methyltransferase. It recognizes 6-O-methylguanine in DNA, removes the methyl group. Transfers the methyl group to an amino acid of the enzyme. "Suicide" mechanism. It is encoded by the ada gene in *E. coli*.

Excision repair

It removes damage (base or DNA backbone) and ss nick/gap provides the 3'OH for DNA Pol I initiation. DNA ligase seals nick.

Nucleotide excision repair

It cuts out a segment of DNA around a damaged base.

Base excision repair

It cuts out the base, then cuts next to the apurinic/apyrimidinic site, and lets DNA Pol I repair.

Mismatch repair

The action of DNA polymerase III (including proofreading exonuclease) results in 1 misincorporation per 10^8 bases synthesized. Mismatch repair reduces this rate to 1 change in every 10^{10} or 10^{11} bases. Recognize mispaired bases in DNA, e.g. G-T or A-C base pairs. These do not cause large distortions in the helix: the mismatch repair system apparently reads the sequence of bases in the DNA.

Role of Methylation in Discriminating Parental and Progeny Strands

dam methylase acts on the A of GATC (note that this sequence is symmetrical or pseudopalindromic). Methylation is delayed for several minutes after replication. Mismatch repair works on the un-methylated strand (which is newly replicated) so that replication errors are removed preferentially.

Action of MutS, MutL, MutH

MutS: recognizes the mismatch (heteroduplex). **MutL**: a dimer; in presence of ATP, binds to MutS-heteroduplex complex to activate MutH. **MutH**:

endonuclease that cleaves 5' to the G in an unmethylated GATC, leaves a nick.

Eukaryotic homologs in mismatch repair

Human homologs to *mutL* (hMLH1) and *mutS* (hMSH2, hMSH1) have been discovered. Mutations in them can cause one of the most common hereditary cancers, hereditary nonpolyposis colon cancer (HNPCC).

Role of umuC and umuD genes in error-prone repair

It is named for the UV **non**-mutable phenotype of mutants with defects in these genes. It is needed for bypass synthesis; the mechanism is under investigation e.g these proteins may reduce the template required for the polymerase. UmuD protein is proteolytically activated by LexA.

Evolution acts on mutations

If we did not have a mutation, then we would all be the same. Then any changes in the environment would be deleterious to all members of the population equally. There would be no evolution but mutations do exist and are supported by a comparison of related organisms.

14. DNA DAMAGE AND REPAIR: BIOLOGICAL INDICATIONS OF DAMAGE TO DNA, EVIDENCE FOR REPAIR SYSTEMS, REPAIR OF THYMINE DIMERS

Introduction

DNA is a relatively stable molecule, but the earth's natural environment is quite toxic, and damage to DNA is inevitable. DNA can also be altered by mistakes made during its own replication or recombination. Damage and sequence alterations to DNA are often quickly repaired, but when they are not, the DNA becomes permanently altered and harbors a mutation. Mutations are changes in the DNA sequence, and when mutations occur in germ-line cells, these changes are inheritable. A cancer cell has mutations that prevent cell death, resulting in loss of cell cycle control and unregulated cell division, which leads to malignant tumors that can end the life of the entire organism. The cell has a limited amount of time to fix the initial alteration and restore the DNA to its normal sequence before replication converts the alteration into a mutation that will be passed on to the next generation.

Mutation

Somatic mutations: It occurs in somatic cells and only affects the individual in which the mutation arises.

Germ-line mutations: They alter gametes and passed to the next generation.

Mutations Are Quantified In Two Ways

- Mutation rate = probability of a particular type of mutation per unit time (or generation).
- Mutation frequency = number of times a particular mutation occurs in a population of cells or individuals.

Types of DNA Mutations

A mutation is a change in a DNA sequence that is propagated through cellular generations. Mutations can be as small as a single base pair or can range from a few base pairs to thousands. Mutations of one or a few base pairs usually result from errors in replication or damaged nucleotides. Mutation can have different effects on gene function fig. 14.1.

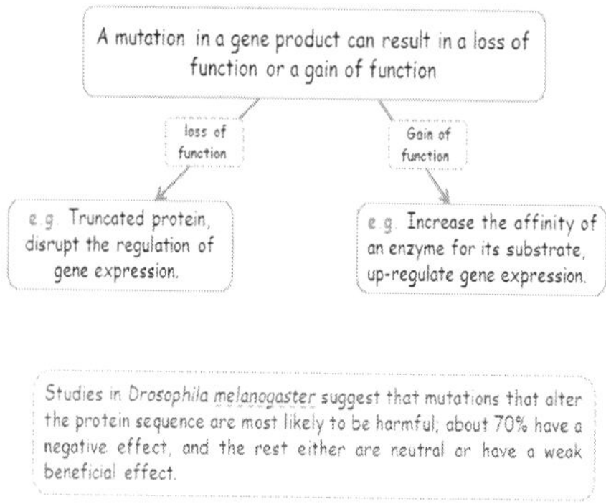

Figure 14.1. Mutation.

A Point Mutation Can Alter One Amino Acid

A change in a single base pair is often referred to as a **point mutation**. Point mutations fall into two categories. A **transition mutation** is the exchange of a purine-pyrimidine base pair for the other purine-pyrimidine base pair: C≡G becomes T=A, or T=A becomes C≡G. A **transversion mutation** is the replacement of a purine-pyrimidine base pair with a pyrimidine-purine base pair or vice versa.

"**For example:** C≡G becomes either G≡C or A=T".

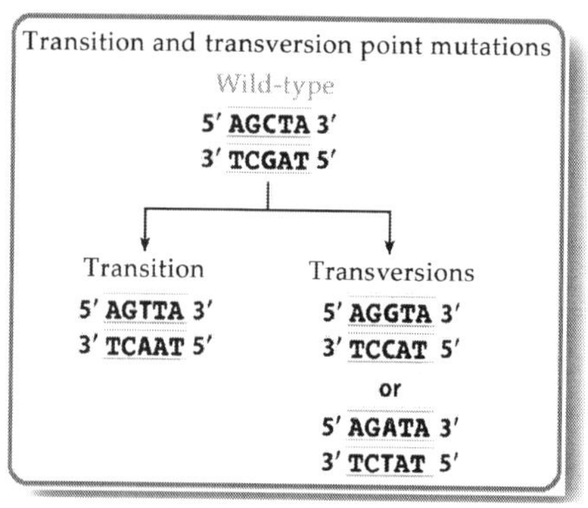

Transition mutations are nearly 10 times more frequent than transversions. A point mutation in the "protein-coding region of a gene can result in an altered protein with partial or complete loss of function. If the protein is central to cell viability, the cell could die. Point mutations are known to cause a wide variety of human diseases. One example is sickle-cell anemia, resulting from a transversion that produces an amino acid change in hemoglobin: a GAG encoding glutamate (E) at residue 6 changes to GTG, encoding valine (V) (E6V). The most harmful mutations are those occurring in the genes involved in DNA repair because these often result in cancer. Many mutations in one cell can result in cancer because a mutation will occur in a gene (or genes) that encodes a protein needed to control cell division. In normal cells, **oncogenes** encode proteins that drive the cell division cycle forward, and **tumor suppressor genes** encode proteins that suppress cell division. Many tumor suppressors are transcription factors that regulate the expression of genes that drive the cell cycle. The transcription factor p53 and the retinoblastoma protein are examples of tumor suppressors that are mutated in many types of cancer.

The DNA sequence encoding a protein is read in codons. Each codon corresponds to an amino acid. A **silent mutation** is a nucleotide change that produces a codon for the same amino acid. For example, GAA and GAG both code for glutamate. A **missense mutation** is a nucleotide change that results in a different amino acid, such as a change from glutamate (GAA) to glutamine (CAA)". A **nonsense mutation** changes the nucleotide sequence so that instead of encoding an amino acid, the triplet functions as a stop codon, terminating the protein. Another type of mutation is the gain or loss of one or more base pairs. **Insertion mutations** occur when one or more base pairs are added to the wild-type sequence. **Deletion mutations** are due to the loss of one or more base pairs. Insertion and deletion mutations are collectively referred to as **indels**. The DNA sequence from the start codon to the stop codon is referred to as a **reading frame**. Because nucleotides are decoded in triplets, an indel mutation of only one or two base pairs in the coding sequence of a protein throws off the reading frame after the mutation, resulting in a **frameshift mutation.**

Mechanisms of DNA Repair

The most prevalent means that cells use to repair damaged DNA is excision repair, of which there are two types: base excision and nucleotide excision repair. "**Base excision repair (BER)** functions at the level of a single damaged nucleotide that distorts DNA very little. It is also the main pathway for the repair of single-strand DNA breaks that lack a ligatable junction and therefore require "cleaning" of the 3′ or 5′ terminus for ligation. **Nucleotide excision repair (NER)** targets large, bulky lesions and removes DNA on either side of them. In contrast to base excision repair, NER does not require specific recognition of a damaged nucleotide and thus it can remove DNA lesions.

Base Excision Repair

(a) In bacteria, a glycosylase excises a damaged nucleotide base, and then an AP endonuclease nicks the backbone at the abasic site. Nick translation by Pol I excise the 5′ deoxyribose phosphate (5′-dRP) and some dNMPs, and synthesizes a new strand. Ligase seals the gap.

(b) Eukaryotic BER, after the first two steps, can take either of two paths. In long patch repair, a DNA polymerase extends the DNA strand from the 3′ terminus, displacing the 5′ single-stranded DNA; this is followed by cleavage by a flap endonuclease and ligation. In short patch repair, only one nucleotide is inserted (by Pol β) prior to ligation.

The NER pathway uses several proteins, including **UvrA**, **UvrB** and **UvrC**, which recognize the lesion and make incisions on either side, allowing UvrD (helicase II) to displace a section of lesion-containing DNA. The single-strand gap is filled in by Pol I, and the DNA is sealed by ligase". A transcription-coupled repair (TCR) path can also be taken in which RNA polymerase stalls at the lesion on the coding strand. After the RNA polymerase is displaced, the reaction proceeds as shown here, using UvrA through UvrD, Pol I, and ligase.

15. REGULATION OF GENE EXPRESSION: BASIC PRINCIPLES OF TRANSCRIPTION AND TRANSCRIPTION IN PROKARYOTES

Prokaryotic Gene Regulation

The DNA of prokaryotes is organized into a circular chromosome supercoiled in the nucleoid region of the cell cytoplasm. „Proteins that are needed for a specific function are encoded together in blocks called **operons**. For example, all of the genes needed to use lactose as an energy source are coded next to each other in the lactose (or lac) operon. In prokaryotic cells, there are three types of regulatory molecules that can affect the expression of operons: repressors, activators, and inducers. **Repressors** are proteins that suppress the transcription of a gene in response to an external stimulus, whereas **activators** are proteins that increase the transcription of a gene in response to an external stimulus. Finally, **inducers** are small molecules that either activate or repress transcription depending on the needs of the cell and the availability of substrate.

Gene Regulation in Prokaryotes

In bacteria and archaea, structural proteins with related functions—such as the genes that encode the enzymes that catalyze the many steps in a single biochemical pathway—are usually encoded together within the genome in a block called an **operon** and are transcribed together under the control of a single **promoter.** This forms a polycistronic transcript. The promoter then has simultaneous control over the regulation of the transcription of these structural genes because they will either all be needed at the same time, or none will be needed. French scientists François **Jacob** (1920–2013) and Jacques **Monod** at the Pasteur Institute were the first to show the organization of bacterial genes into operons, through their studies on the ***lac* operon** of *E. coli*. They found that in *E. coli*, all of the structural genes that encode enzymes needed to use lactose as an energy source lie next to each other in the lactose (or *lac*) operon under the control of a single promoter, the *lac* promoter. For this work, they won the Nobel Prize in Physiology or Medicine in 1965.

Although eukaryotic genes are not organized into operons, prokaryotic operons are excellent models for learning about gene regulation generally. There are some gene clusters in eukaryotes that function similarly to operons. Many of the principles can be applied to eukaryotic systems and contribute to our understanding of changes in gene expression in eukaryotes that can result in pathological changes such as cancer. Each operon includes DNA sequences that influence its own transcription; these are located in a region called the regulatory region. The regulatory region includes the promoter and the region surrounding the promoter, to which **transcription factors**, proteins encoded by regulatory genes, can bind. Transcription factors influence the binding of **RNA polymerase** to the promoter and allow its progression to transcribe structural genes. A **repressor** is a transcription factor that suppresses transcription of a gene in response to an external stimulus by binding to a DNA sequence within the regulatory region called the **operator**, which is located between the RNA polymerase binding site of the promoter and the transcriptional start site of the first structural gene. Repressor binding physically blocks RNA polymerase from transcribing structural genes. Conversely, an **activator** is a transcription factor that increases the transcription of a gene in response to an external stimulus by facilitating RNA polymerase binding to the promoter. An **inducer**, a third type of regulatory molecule, is a small molecule that either activates or represses transcription by interacting with a repressor or an activator. In prokaryotes, there are examples of operons whose gene products are required rather consistently and whose expression, therefore, is unregulated. Such operons are constitutively expressed" meaning they are transcribed and translated continuously to provide the cell with constant intermediate levels of the protein products. Such genes encode enzymes involved in housekeeping functions required for cellular maintenance, including DNA replication, repair, and expression, as well as enzymes involved in core metabolism. In contrast, there are other prokaryotic operons that are expressed only when needed and are regulated by repressors, activators, and inducers".

The *trp* Operon: A Repressor Operon

Bacteria such as *E. coli* need amino acids to survive. «**Tryptophan** is one such amino acid that *E. coli* can ingest from the environment. *E. coli* can also synthesize tryptophan using enzymes that are encoded by five genes. These five genes are next to each other in what is called the tryptophan (*trp*) operon . If tryptophan is present in the environment, then *E. coli* does not need to synthesize it and the switch controlling the activation of the genes in the *trp* operon is switched off. However, when tryptophan availability is low, the switch controlling the operon is turned on, transcription is initiated, the genes are expressed, and tryptophan is synthesized. A DNA sequence that codes for proteins is referred to as the coding region. The five coding regions for the tryptophan biosynthesis enzymes are arranged sequentially on the chromosome in the operon. Just before the coding region is the transcriptional start site. This is the region of DNA to which RNA polymerase binds to initiate transcription. The promoter sequence is upstream of the transcriptional start site; each operon has a sequence within or near the promoter to which proteins (activators or repressors) can bind and regulate transcription. A DNA sequence called the operator sequence is encoded between the promoter region and the first *trp* coding gene. This operator contains the DNA code to which the repressor protein can bind. When tryptophan is present in the cell, two tryptophan molecules bind to the *trp* repressor, which changes shape to bind to the *trp* operator. The binding of the tryptophan–repressor complex at the operator physically prevents the RNA polymerase from binding, and transcribing the downstream genes. When tryptophan is not present in the cell, the repressor by itself does not bind to the operator; therefore, the operon is active and tryptophan is synthesized. Because the repressor protein actively binds to the operator to keep the genes turned off, the *trp* operon is negatively regulated and the proteins that bind to the operator to silence *trp* expression are negative regulators.

Catabolite Activator Protein (CAP): An Activator Regulator

Just as the *trp* operon is negatively regulated by tryptophan molecules, there are proteins that bind to the operator sequences that act as a positive regulator to turn genes on and activate them. For example, when glucose is scarce, *E. coli* bacteria can turn to other sugar sources for fuel. To do this, new genes to process these alternate sugars must be transcribed. When glucose levels drop, cyclic AMP (cAMP) begins to accumulate in the cell. The cAMP molecule is a signaling molecule that is involved in glucose and energy metabolism in *E. coli*. When glucose levels decline in the cell, accumulating cAMP binds to the positive regulator catabolite activator protein (CAP), a protein that binds to the promoters of operons that control the processing of alternative sugars. When cAMP binds to CAP, the complex binds to the promoter region of the genes that are needed to use the alternate sugar sources (Figure 1). In these operons, a CAP binding site is located upstream of the RNA polymerase binding site in the promoter. This increases the binding ability of RNA polymerase to the promoter region and the transcription of the genes.

lac Operon: An Inducer Operon

The third type of gene regulation in prokaryotic cells occurs through **inducible operons**, which have proteins that bind to activate or repress transcription depending on the local environment and the needs of the cell. The *lac* operon is a typical inducible operon. As mentioned previously, *E. coli* is able to use other sugars as energy sources when glucose concentrations are low. To do so, the cAMP–CAP protein complex serves as a positive regulator to induce transcription. One such sugar source is lactose. The ***lac* operon** encodes the genes necessary to acquire and process the lactose from the local environment. CAP binds to the operator sequence upstream of the promoter that initiates transcription of the *lac* operon. However, for the *lac* operon to be activated, two conditions must be met. First, the level of glucose must be very low or non-existent. Second, lactose must be present. Only when glucose is absent and lactose is present will the *lac* operon be transcribed. This makes sense for the cell, because it would be energetically wasteful to create the proteins to process lactose if glucose was plentiful or lactose was not available. If glucose is absent, then CAP can bind to the operator sequence to activate transcription. If lactose is absent, then the repressor binds to the operator to prevent transcription. If either of these requirements is met, then transcription remains off. Only when both conditions are satisfied is the *lac* operon transcribed

Prokaryotic Gene Regulation

As we've just learned, there are three types of regulatory molecules that can affect the expression of operons: repressors, activators, and inducers. **Repressors** are proteins that suppress the transcription of a gene in response to an external stimulus. In other words, a repressor keeps a gene "off.". **Activators** are proteins that increase the transcription of a gene in response to an external stimulus. In other words, an activator turns a gene "on." **Inducers** are small molecules that either activate or repress transcription depending on the needs of the cell and the availability of substrate. Inducers basically help speed up or slow down "on" or "off" by binding to a repressor or activator.

16. TRANSCRIPTION IN PROKARYOTES TRANSCRIPTION IN EUKARYOTES, ITS REGULATION, RNA PROCESSING AND RNPS– MULTIPLE RNA POLYMERASES, SIGMA LIKE FACTORS IN EUKARYOTES; HETEROGENEOUS NUCLEAR RNA; MESSENGER RNA - STRUCTURE AND COMPLEXITY; INTERRUPTED GENES AND RNA SPLICING; EXPRESSION OF SPECIFIC GENES; GENES FOR RIBOSOMAL RNA; HISTONE GENES; GLOBIN GENES; HEAT-SHOCK GENES

Introduction

It is the process of the formation of the RNA. It takes place through the usual process of complementary base pairing, catalyzed and scrutinized by the enzyme RNA polymerase. It occurs unidirectionally in which RNA chain (transcript) is synthesized from 5' to 3 directions. It is chemically and enzymatically very similar to DNA replication. However, there are the following important that differ from that of replication. The new strand is made from ribonucleotides (NTPs) rather than deoxyribonucleotides (NTPs). It is primer independent. RNA polymerase can initiate a new RNA chain on a DNA template without the need for a primer. It is a selective process i.e transcription selectively copies only certain parts of the genome called a transcription unit. In contrast, during replication entire genome replicates. Only one of the DNA strands acts as a template on which the RNA strand is built. It is unidirectional.

Prokaryotes

Promoter

Pribnow box (also called the -10 element) – TATAAT and 35 element-TTGACA.

Eukaryotes

The basic core promoter is the TATA box (TATAAA(A)) within 50 bp upstream of the start site which found in unicellular eukaryotes. The core promoter plus has downstream promoters and proximal promoter elements.

Initiator sequences

Regulatory elements/Response element: **Response elements** are the recognition sites of certain transcription factors. Most of them are located within 1 kb from the transcriptional start site. Enhancer elements are-upon binding with transcription factors (**activators**) and they can enhance transcription which are located either upstream or downstream of the transcriptional initiation site.

- Upstream enhancer elements
- Downstream enhancers
- Distal enhancer elements

Silencers are upon binding with transcription factors (**repressors**) and can repress transcription.

Proteins Involved in Transcription

RNA polymerase transcription factors are TFIIA, TFIIB, TFIID, TFIIE, TFIIF, TFIIH. transcription factors that bind to regulatory elements are holoenzyme or initiation complex and transcription factors have common DNA binding motifs fig. 16.1.

- Zinc finger
- Helix-turn-helix
- Leucine zipper

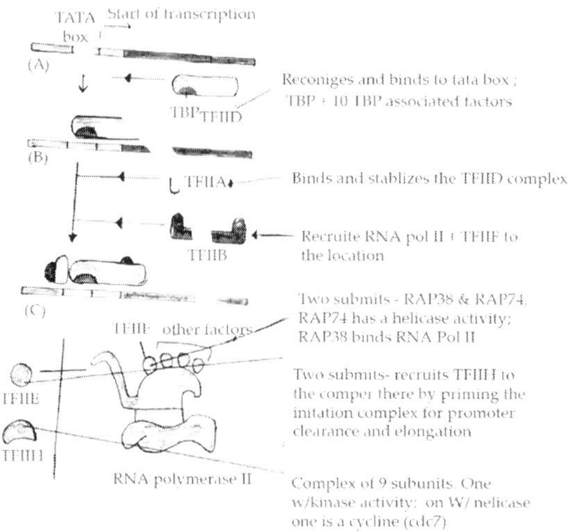

Figure 16.1. Transcription factors that bind to regulatory elements.

Synthesis of enzymes comes into the control of genetic material i.e., DNA in living organisms. Enzymes synthesis occurs by two processes such as translation and transcription. Translation refers to the synthesis of mRNA. It is controlled by the promoter region of the gene.

Pathway for Gene Expression

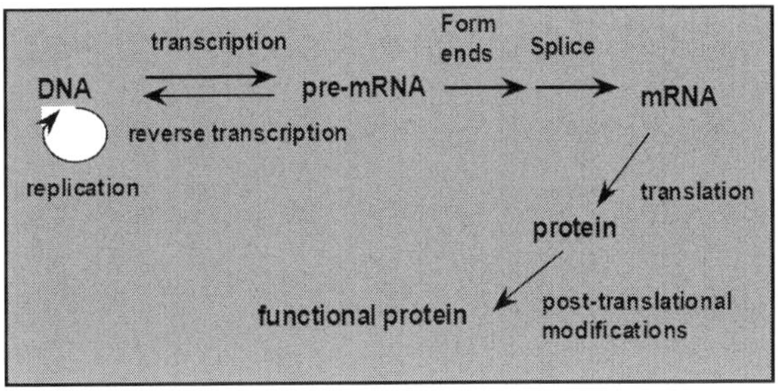

Figure 16.2. Central dogma.

The Reaction Catalyzed By RNA Polymerase

It catalyzes the synthesis of RNA directed by DNA as a template = **transcription.** Makes an RNA chain with a sequence complementary to the template strand of DNA. It can start RNA synthesis at a site on the DNA template *E. coli* RNA polymerase. It synthesizes all classes of RNA.

- mRNA
- rRNA
- tRNA

Core= α_2**ßß' catalyzes the elongation of an RNA chain.** Holoenzyme = α_2**ßß's catalyzes initiation or RNA synthesis specifically at a promoter.**

Promoter

The sequence of DNA is required for accurate, specific initiation of transcription. The sequence of DNA to which RNA polymerase binds to accurately initiate transcription. In most cases, RNA polymerase binds to a DNA sequence including the initiation site, but it can be directed there by sequences flanking the initiation site.

RNA Polymerase At A Promoter

- **α**: assembly and binds to UP.
- **ß+ß'**: form catalytic center.
- 6: binds -10 and -35 of the promoter to confer specificity during initiation.

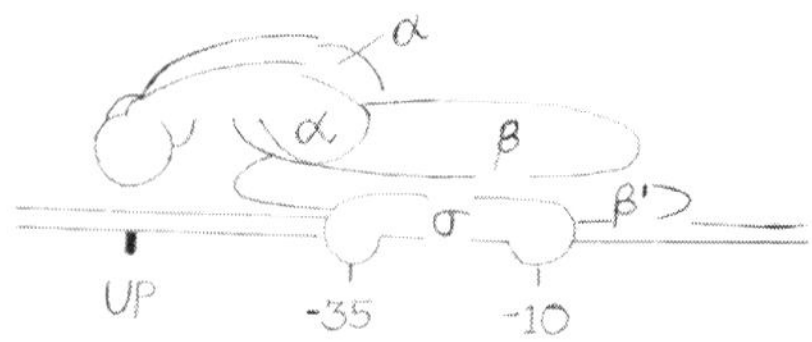

Mode of Action of 6 Factors

The 6 factor causes RNA polymerase to be selective in its choice of initiation sites by affecting the dissociation rate of polymerase from DNA. The core dissociates from general DNA with a half-time of 60 min and functions in elongation. Holoenzyme dissociates from general DNA with a half-time of 1 sec. Holoenzyme dissociates from promoter DNA with a half-time of hours (in absence of rNTPs).

Transcription Cycle

Initiation: Holoenzyme binds to the promoter, unwinds DNA, and forms phosphodiester bonds between 7 to 12 nucleotides.

Elongation: The core elongates RNA with high processivity and it may use NusA.

Termination: Polymerase dissociates from template DNA and releases new RNA fig. 16.3.

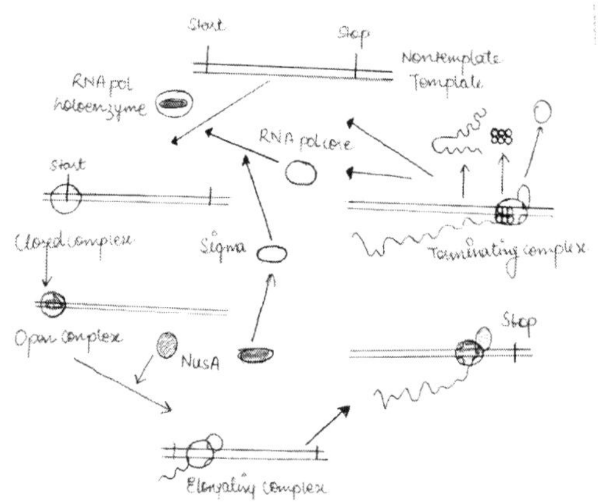

Figure 16.3. Termination.

Sites on RNA Polymerase Core

Enzyme covers about 60 bp of DNA, with about 17 bp unwound = transcription bubble. The bubble must contact the active site for polymerization. At the beginning of the bubble, the DNA is unwound, implicating a helicase activity. At the end of the bubble, the DNA is rewound. Effects of transcription on supercoiling of the template. The unwinding (untwisting) and rewinding of the DNA template introduce positive supercoils ahead of the polymerase and negative supercoils behind it. Unwinding will decrease T by 1 for every 10 bp unwound. Since DL=0, DW=-DT, and DW=+1 for every 10 bp unwound. The opposite occurs upon rewinding. DT=+1 for every 10 bp rewound, and thus DW=-1 for every 10 bp rewound. Topoisomerases aid transcription in vivo fig. 16.4.

RNA Polymerases in Eukaryotes

Name	Makes	α-amanitin
RNA Polymerase I	pre-rRNA	insensitive
RNA Polymerase II	pre-mRNA some snRNAs	very sensitive
RNA Polymerase III	pre-tRNA other small RNAs some snRNAs	less sensitive

Figure 16.4. RNA Polymerase.

Subunit Structure of Eukaryotic RNA Polymerases

All 3 have multiple subunits (8 to 14). The molecular weight of each polymerase is about 500,000. Some subunits are common to all 3 RNA polymerases. All 3 RNA polymerases have subunits that are homologous to the bacterial ß, ß' and α subunits. The transcription and translation take place in different cellular comparts: transcription takes place in the nucleus and translation takes place in the cytosol. In prokaryotes, transcription of mRNA and translation occur simultaneously. Thus, mRNA molecules undergo little or no modification after synthesis by RNA polymerase in prokaryotes. In contrast "pre-tRNA and pre-rRNA undergo processing like cleavage, the addition of nucleotides and chemical modification after synthesis. Although both prokaryotes and eukaryotes modify pre-tRNA and pre-rRNA, eukaryotes very extensively process pre-mRNA destined to become mRNA. The primary transcript of an RNA polymerase is referred to as pre-RNA. Processing of eukaryotic pre-mRNA involves 5' capping, 3' polyadenylation, splicing and RNA editing before being transported to the cytosol" where they are translated by ribosomes". The final product of gene expression is a polypeptide chain of amino acids whose sequence was prescribed by the genetic code.

- mRNA transcribed from genomic DNA.
- Transfer RNA (tRNA) to transport amino acids.
- Ribosome to "read" mRNA, align amino acids attached to tRNA and create the peptide bonds between adjacent amino acids.

Ribosome

There are two subunits which are composed of protein and ribosomal RNA (rRNA). The rRNA is a structural component of the ribosome subunits fig.16.5.

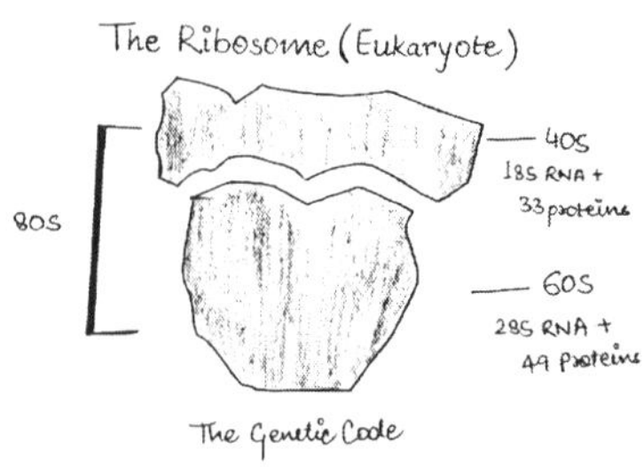

Figure 16.5. Ribosome.

tRNA

tRNAs are small, extremely stable RNA structures shaped like a cloverleaf due to internal base pairing. They are almost identical in both prokaryotes and eukaryotes. They contain a number of post-transcriptional modifications, including non-traditional bases (other than (A, U, G and C). The amino acid attaches to the 3' end of the tRNA. tRNA can base pair with the codon in mRNA at the **anticodon** on the tRNA.

Charging tRNA With an Amino Acid

An individual tRNA is specific for a single amino acid. (There are at least 20 different tRNAs). Charging is the process of covalently attaching an amino acid to the tRNA. Charging is accomplished by aminoacyl tRNA synthetases, each of which is specific for an amino acid.

- **Initiation:** the two subunits of the ribosome come together and the start codon on the mRNA in the ribosome is aligned to set the reading frame.
- **Elongation:** charged tRNAs attach and peptide bonds form between the amino acids.
- Termination: It is the process where a nascent RNA is released from its complex with RNA polymerase and the DNA template.

Prokaryotes

Translation takes place directly after transcription. mRNA is not modified. Transcription and translation take place in the same area.

Eukaryotes

The transcript is modified before leaving the nucleus (5' cap and 3' poly-A tail).

- Modifications increase the translation efficiency and lifespan of the mRNA.
- Translation takes place on ribosomes located in the rough endoplasmic reticulum (translation is physically separated from transcription).
- Influence on gene expression forms the basis of function in many antibiotics.

- The purpose is to impair function in the prokaryote without disrupting function in eukaryotes.

Antibiotics

- **Tetracycline:** Blocks attachment of the AA-tRNA to the A site in the ribosome.
- **Erythromycin:** Blocks the movement of the ribosome down the mRNA.
- **Streptomycin:** Blocks the formation of the initiation complex and causes miscoding.
- **Rifamycin:** Blocks transcription by blocking the binding of RNA polymerase.

Genetic Code

It consists of 64 triplet codons (A, G, C, U) $4^3 = 64$). All codons are used in protein synthesis. It has 20 amino acids. It has 3 termination (stop) codons: UAA, UAG, and UGA. So, AUG (methionine) is the start codon (also used internally). It has multiple codons for a single amino acid = degeneracy. It has 5 amino acids are specified by the first two nucleotides only.

The Genetic Code

UUU UUC Phenylalanine UUA UUG Leucine	UCU UCC UCA UCG Serine	UAU UAC Tyrosine UAA UAG Stop	UGU UGC Cysteine UGA Stop UGG Tryptophan
CUU CUC CUA CUG Leucine	CCU CCC CCA CCG Proline	CAU CAC Histidine CAA CAG Glutamine	CGU CGC CGA CGG Arginine
AUU AUC AUA Isoleucine AUG Methionine	ACU ACC ACA ACG Threonine	AAU AAC Asparagine AAA AAG Lysine	AGU AGC Serine AGA AGG Arginine
GUU GUC GUA GUG Valine	GCU GCC GCA GCG Alanine	GAU GAC Aspartic acid GAA GAG Glutamic acid	GGU GGC GGA GGG Glycine

Codon-Anticodon Interactions

The codon-anticodon base-pairing is antiparallel. The third position in the codon is frequently degenerate. One tRNA can interact with more than one codon (therefore 50 tRNAs). The Wobble rules:

- C with G or I (inosine).

- A with U or I.
- G with C or U.
- U with A, G, or I.
- I with C, U, or A.

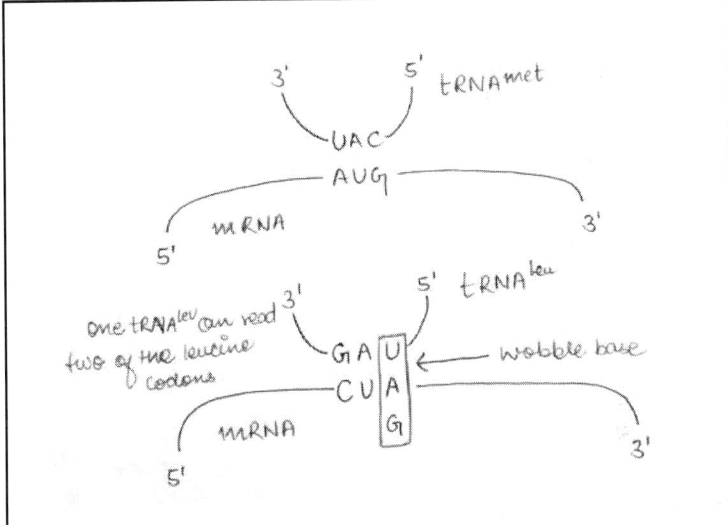

Termination Point

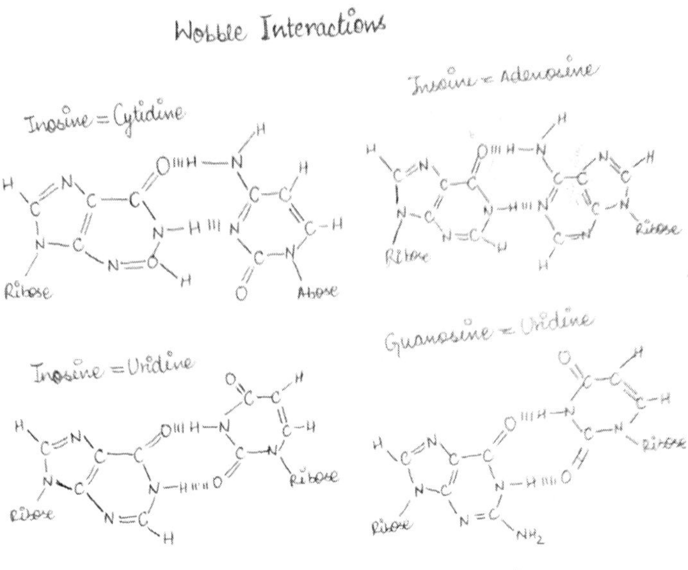

Initiation in Prokaryotes and Eukaryotes

Initiation can occur at internal AUG codons in prokaryotic mRNA. Initiation in eukaryotes occurs only at the first AUG codon. Lac operon in *E. coli* is transcribed as a polycistronic mRNA with multiple AUG codons.

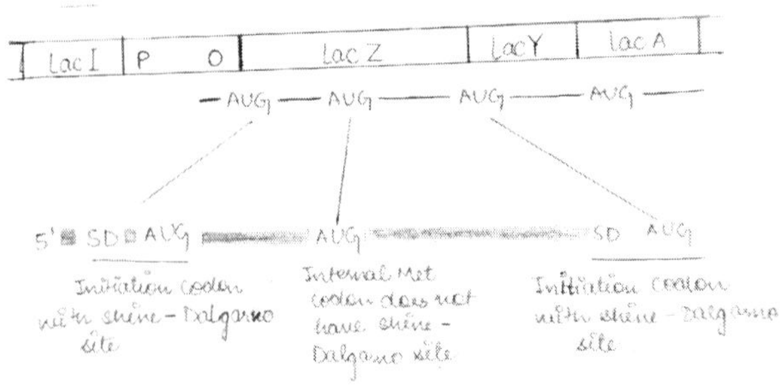

Reading Frame

The reading frame is determined by the AUG initiation codon. Every subsequent triplet is read as a codon until reaching a stop codon.

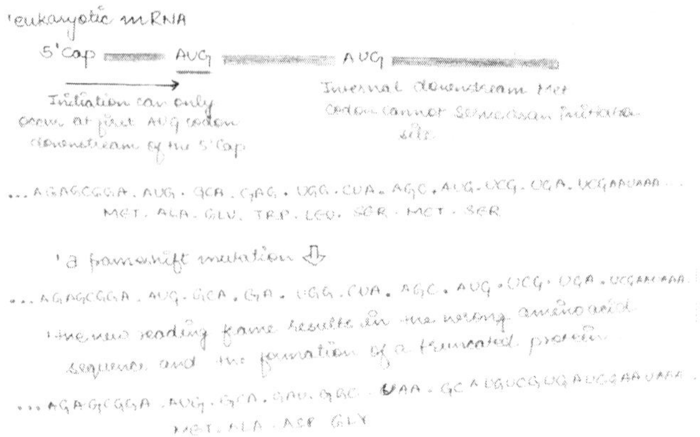

Mutations Affecting Translation

Hemoglobin wayne (3' terminal frameshift mutation). Missense mutations are (e.g., AGC Ser to AGA Arg). Nonsense mutations are (e.g., UGG Trp to UGA Stop). Read through, reverse terminator, or sense mutations (e.g., UAA Stop to CAA Gln) as in hemoglobin constant spring. Silent mutations are (e.g., CUA Leu to CUG Leu) do not affect translation.

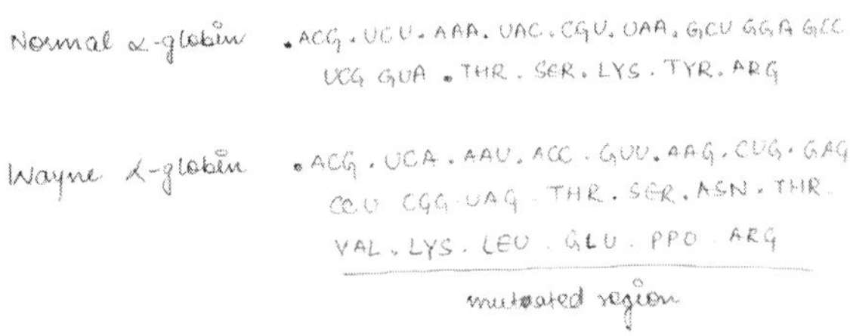

Post-Translational Modifications

- Cleavage of signal peptides
- Phosphorylation
- Amidation
- Glycosylation
- Hydroxylation
- Ubiquitination
- Addition of prosthetic groups
- Iodination
- Adenylation
- Sulfonation
- Prenylation
- Myristoylation
- Acylation
- Acetylation
- Methylation
- Oxidative crosslinking
- N-Glutamyl cyclization
- Carboxylation

N-terminal Modifications

- N-formyl Met (by-product of the translation process and incomplete cleavage by deformylase).
- Aminopeptidase cleavage of Met.
- N-terminal acetylation.
- N-terminal myristoylation.

- N-terminal glutamyl cyclization.

C-terminal Modification

- C-terminal prenylation
  - Farnesylation
  - Geranylation/geranyl-geranylation
- C-terminal amidation

Phosphorylation

- It has hydroxyl groups - Ser, Thr, and Tyr.
- It has amide groups of His, Lys & carboxy group Asp.
- It often causes changes in protein conformation and/or protein-protein interactions.

C-terminal Amidation

- Neuropeptides and hormones
- C-terminal glycine is hydroxylated alpha-hydroxy-glycine.
- Glyoxylate. C-terminal amidation is essential to the biological activity of many neuropeptides and hormones.

Hydroxylation and Carboxylation

- Hydroxylation (previous lecture).
  - Lys and Pro (e.g. collagen).
  - Requires vitamin C.
- Carboxylation
  - Requires vitamin K.

Iodination

- Special case: Tyr $\rightarrow$ thyroxin.

Protein Acylation

- Methylation and acetylation.
- Acetylation of histone Lys.
- S-acylation (Cys).

- Prenylation (C-terminus).
- Myristoylation (N-terminus).

Acetylation

N-terminal acetylation

It has no simple sequence consensus. It is usually accompanied by cleavage of the N-terminal Met.

Methylation

It can occur at carboxyls, amides, sulfhydryl, etc. S-Adenosyl Methionine (S-AM) is the activated carbon source.

Histones

DNA binding proteins form nucleosomes by binding four types of histones. It affects gene expression.

Operon

A unit of bacterial gene expression and regulation, including structural genes and control elements in DNA recognized by regulator gene product(s). Trans-acting: A product that can function on any copy of its target DNA. This implies that it is a diffusible protein or RNA. Cis-acting – A site that affects the activity only of sequences on its own molecule of DNA (or RNA); this property usually implies that the site does not code for protein. **Regulator gene**: A gene that codes for a product that controls the expression of other genes (usually at the level of transcription). **Structural gene:** A gene that codes for any RNA or protein product other than a regulator. In negative regulation, a repressor protein binds to an operator to prevent a gene from being expressed. In positive regulation, a transcription factor is required to bind at the promoter in order to enable RNA polymerase to initiate transcription.

- In inducible regulation, the gene is regulated by the presence of its substrate (the **inducer**).
- In repressible regulation, the gene is regulated by the product of its enzyme pathway (the **corepressor**).

Gene regulation *in vivo* can utilize any of these mechanisms, resulting in all four combinations: negative inducible, negative repressible, positive inducible and positive repressible. Genes coding for proteins that function in the same pathway may be located adjacent to one another and controlled as a single unit that is transcribed into a polycistronic mRNA.

The lac operon is negative inducible

Transcription of the *lacZYA* operon is controlled by a repressor protein (the *lac* repressor) that binds to an operator that overlaps the promoter at the start of the cluster. The constitutive expression-A state in which a gene is expressed continuously. In the absence of β-galactosides, the *lac* operon is expressed only at a very low (basal) level. The repressor protein is a tetramer of identical subunits coded by the *lacI* gene. **β**-galactoside sugars, the substrates of the *lac* operon, are its inducer. The addition of specific β-galactosides induces transcription of all three genes of the *lac* operon. The *lac* mRNA is extremely unstable; as a result, induction can be rapidly reversed.

Lac repressor is controlled by a small-molecule inducer

An inducer functions by converting the repressor protein into a form with a lower operator affinity. Repressor has two binding sites, one for the operator DNA and another for the inducer. **Gratuitous inducer:** Inducers that resemble authentic inducers of transcription, but are not substrates for the induced enzymes.

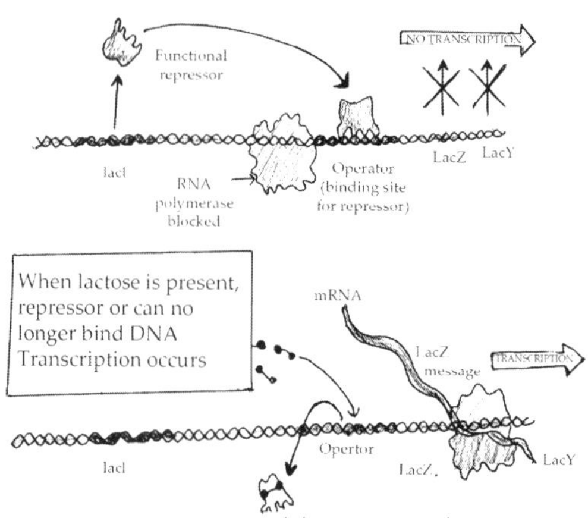

Lac operon

The repressor is inactivated by an allosteric interaction in which binding of the inducer at its site changes the properties of the DNA-binding site (allosteric control). The true inducer is allolactose, not the actual substrate of β-galactosidase.

cis-Acting constitutive mutations identify the operator

Mutations in the operator cause constitutive expression of all three *lac* structural genes. These mutations are *cis*-acting and affect only those genes on the contiguous stretch of DNA._Mutations in the promoter prevent the expression of *lacZYA* and are uninducible and *cis*-acting. **Cis-dominant:** A site or mutation that affects the properties only of its own molecule of DNA, often indicating that a site does not code for a diffusible product.

Trans-acting mutations identify the regulator gene

Mutations in the *lacI* gene are Trans-acting and affect the expression of all *lacZYA* clusters in the bacterium. "Mutations that eliminate *lacI* function cause constitutive expression and are recessive (*lacI*$^{-}$). Mutations in the DNA-binding site of the repressor are constitutive because the repressor cannot bind the operator. Mutations in the inducer-binding site of the repressor prevent it from being inactivated and cause uninducibility. When mutant and wild-type subunits are present, a single *lacI*$^{-d}$ mutant subunit can inactivate a tetramer whose other subunits are wild-type. It is a dominant negative. Interallelic complementation – The change in the properties of a heteromultimeric protein brought about by the interaction of subunits coded by two different mutant alleles. The mixed protein may be more or less active than the protein consisting of subunits of only one or the other type. Negative complementation – This occurs when interallelic complementation allows a mutant subunit to suppress the activity of a wild-type subunit in a multimeric protein. *lacI*$^{-d}$ mutations occur in the DNA-binding site". Their effect is explained by the fact that repressor activity requires all DNA-binding sites in the tetramer to be active.

Positive Gene Regulation

If glucose levels are low (along with overall energy levels), then cyclic AMP (cAMP) binds to cAMP receptor protein (CRP) which activates transcription. If glucose levels are sufficient and cAMP levels are low (lots of ATP), then the CRP protein has an inactive shape and cannot bind upstream of the *lac* promoter.

Points of Control of Gene Regulation

The control of gene expression can occur at any step in the pathway from a gene to a functional protein.

1. Packing/unpacking DNA
2. Transcription
3. mRNA processing
4. mRNA transport
5. Translation
6. Protein processing
7. Protein degradation

17. THE DYNAMIC GENOME: MOBILE GENETIC ELEMENTS IN EUKARYOTES (JUMPING GENES)

Introduction

DNA sequence that can move in the genome. It is also called a mobile DNA element or transposable element. It has "selfish DNA" that exist only to maintain themselves.

Transposition: The process by which these sequences are copied and inserted into a new site in the genome. It has probably had a significant influence on evolution. In the 1940s, Barbara McClintock discovered the first transposable element in maize, earned a Nobel prize in 1983. In the late 1960s, transposition was also found in bacteria.

Retrotransposons

Most mobile elements in bacteria are DNA transposons. In contrast, most mobile elements in eukaryotes are retrotransposons, but eukaryotic DNA transposons also occur.

DNA Transposons

- Bacterial insertion sequences (IS element).
- P element in *Drosophila.*

Retrotransposons

- LTR retrotransposons.
- Non-LTR retrotransposons: the most common type of transposons in mammals.

Non-LTR Retrotransposons

- Long interspersed elements (LINEs) ≈6 kb in humans account for **21%** of the genome.
- Short interspersed elements (SINEs) ≈300 bp in humans account for **13%** of the genome.

Transposable Elements

Transposable elements are called mobile genetic elements. It comprises

45% of human chromosomal DNA "middle repetitive DNA". It contributes to spontaneous mutation, genetic rearrangements, and horizontal transfer of genetic material. The aid speciation and genomic change (in bacteria transposons are often associated with antibiotic resistance genes). The cells must depress transposition to insure genetic stability.

Types of Transposable Elements

Composite Bacterial Transposons

It has repeated ends, usually inverted, sometimes direct. The repeated ends themselves are IS elements and can independently transpose. The ends mobilize all intervening DNA. It often contains antibiotic resistance genes (examples Tn3 (ampicillin), Tn5 (kanamycin), Tn10 (tetracycline). It often resides on plasmids.

Types of DNA Transposons

Excisive mechanism examples: Tn5, Tn10, P elements.

Replicative mechanism examples: Tn3, bacteriophage Mu.

Excisive Transposons

They have a cut-and-paste mechanism and are called excisive transposons. It has cut them out of the original site, producing double-strand break. It is cut target site and ligate to element ends, thereby inserting at the new site. The original site breaks repaired usually with sister chromosome, restoring transposon at the original site and sometimes end healed without transposon, can also be associated with deletion at the excision site.

Replicative Transposons

The original cut of the transposon is the only nick and only one strand is initially ligated. The element replicates itself. It produces as intermediate a "co-integrate" structure. It has co-integrate is resolved by resolvase (as TnpR of Tn3) and at a specific site (as *res* of Tn3).

Degenerate Transposons

Many naturally occurring transposable elements have suffered from mutations and are no longer active. Some of these may have cis-acting end mutations and cannot be mobilized. Others may have intact ends but no transposase:

these can be mobilized by an element that is tnp+ ("autonomous" element). Ac Ds system is an example of the latter: Ac (activator) can mobilize Ds (dissociator).

- MITEs (minature inverted repeat transposable elements) are nonautonomous DNA elements.
- SINEs are retrotransposon version (LINEs).

Classification of Retroelements

It has an obligate RNA intermediate, use reverse transcriptase (RT, RNA-dependent DNA polymerase).

- LTR-retroelements: long terminal repeatsTy1/copia, Ty3/gypsy and retroviruses.
- Non-LTR-retroelements "retroposons" LINES.

Characteristics of LTR Retroelements

Long terminal repeats are required for the replication cycle. **Genes:** gag, pol, (viruses also have env). Polymerase is a polyprotein that gives rise to reverse transcriptase, in (integrase) RH (RNase H), protease. It has forms VLPs virus-like particles. Integrase is functionally and structurally similar to transposase of DNA transposons, DDE motif. The integration gives characteristic TSD.

Characteristics of Non-LTR Retroelements

It contains 2 ORFs, ORF1, ORF2. The variable TSD. ORF2 gives rise to endonuclease, (similar to APE) and RT. It uses target primed reverse transcription TPRT. It can transduce 3' downstream non-element segments.

Human L1 (LINE-1) Retroelement

It has presents 15% of human DNA. It contains 520,000 copies, only 3-5,000 are full-length. It is associated with human disease loci. Transpose specifically in the germline.

18. BRIEF INTRODUCTION TO THE COMPLEXITY OF THE EUKARYOTIC GENOME

Introduction

By the 1960s, when **Roy Britten and Eric Davidson** were studying eukaryotic gene regulation, they knew that there was more than enough DNA to account for the genes needed to encode an organism. It was also likely that DNA was more structurally complex than originally thought. They knew that "cesium chloride (CsCl) density gradient centrifugation separated molecules based on differences in density and that fragmented DNA would separate into a main and a minor band of different density in the centrifuge tube. The minor band was dubbed satellite DNA, recalling the *Sputnik* satellite recently launched by Russia. DNA bands of different densities could not exist if the proportions of A, G, T and C in DNA (already known to be species-specific) were the same throughout a genome. Instead, there must be regions of DNA that are richer in A-T than G-C pairs and vice versa. Analysis of satellite bands that moved further on the gradient (i.e., were denser) than the main band were indeed richer in GC content. Those that lay above the main band were more AT-rich. Consider early estimates of how many genes it might take to make a human, mouse, chicken or petunia: about 100,000! We know now that it takes fewer! Nevertheless, even with inflated estimates of the number of genes it takes to make a typical eukaryote, their genomes contain 100-1000 times more DNA than necessary to account for 100,000 genes.

How then to explain this extra DNA? Britten and Davidson's elegant renaturation kinetics experiments revealed some physical characteristics of genes and so-called 'extra' DNA. The first step in a renaturation kinetic experiment is to shear DNA isolates to an average size of 10 Kbp by pushing high molecular weight DNA through a hypodermic needle at constant pressure. The resulting double-stranded fragments (dsDNA) are then heated to 100°C to denature (separate) the two strands. The solutions are then cooled to 600C to allow the single-stranded DNA (ssDNA) fragments to slowly re-form complementary double strands". At different times after incubation at 60°C, the partially renatured DNA was sampled and ssDNA and dsDNA were separated and quantified. Britten and Davidson defined the relative amounts of repeated and unique (or single copy) DNA sequences in an organism's genome as its genomic complexity. Thus, prokaryotic genomes have lower genomic complexity than eukaryotes.

Chapter 3

PLANT METABOLIC PATHWAYS

19. INTRODUCTION TO METABOLISM: ANABOLISM AND CATABOLISM, METABOLIC COMPARTMENTS, THERMODYNAMICS, TRANSPORT SYSTEMS, TRANSLOCATORS

Introduction

A metabolic pathway has many steps that begin with a specific molecule and end with a product that are each catalyzed by a specific enzyme. They are two metabolic pathways fig 19.1.

Catabolic Pathways

It breaks down complex molecules into simpler compounds eg. amylase breaks complex starches into simple sugars and it is the process of cellular respiration.

Anabolic Pathways

It consumes energy to build complicated molecules eg. The anabolic steroids build the muscle. The building of a protein from amino acids.

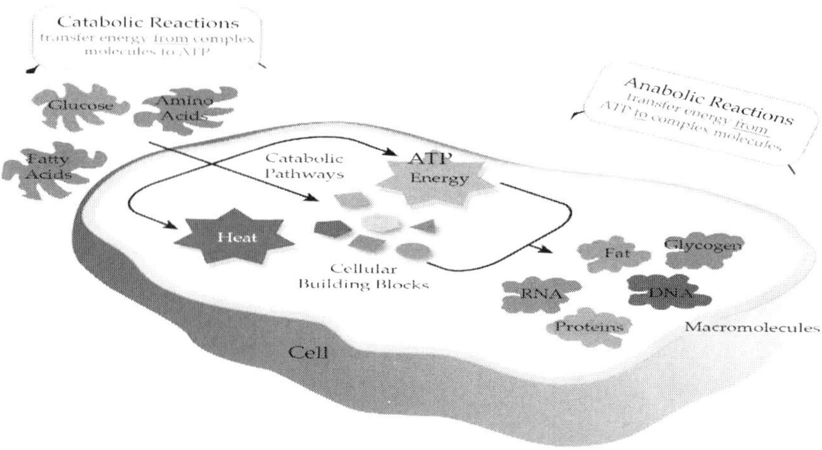

Figure 19.1. Metabolism.

Different Types of Energy (energy=the ability to do work or cause change)

Potential

The energy stored in an object. It is measured in joules and chemical energy is the potential energy of a chemical reaction.

Kinetic

The energy of an object in motion. It is measured in joules. Thermal energy is the kinetic energy of atoms or molecules.

Laws of Thermodynamics

1st Law: Energy can be transferred and transformed but it can't be created or destroyed.

2nd Law: Every energy transfer or transformation increases the entropy of the universe.

Thermodynamics is the metabolic reactions in organisms. It must occur at body temperature. The body temperature does not get substrates to their transition state. The active site of enzymes lowers the amount of energy needed to reach a transition state.

Mechanism Of Transport

Nonmediated transport: It has through simple diffusion.

Mediated transport: It has required specific carriers.

Passive mediated transport: Specific molecules which flow from high to low concentration. The net flow will cease when the concentration gradient is no longer in existence.

Active transport: It is from low to high concentration and endergonic processes. It must be coupled to a sufficiently energetic (exergonic) process to make it possible.

Classification

Uniport is **the** movement of a single type of molecule at a time.

Symport is **the** simultaneous movement of two different molecules in the same direction.

Antiport is **the** simultaneous movement of two different molecules in opposite directions.

Nonmediated transport is through simple diffusion, driven by (electro) chemical potential gradient. The substance diffuses in the direction that eliminates its concentration gradient. The rate of diffusion is: (a) proportional to the magnitude of the (electro)chemical gradient [For many nonelectrolytes, their fluxes across erythrocyte membranes increase linearly with $(A)_{out}$ $-(A)_{in}$ (b) dependent on the solubility of the substance in the membrane's nonpolar core.

Mediated transport is through the action of specific proteins or other molecules referred to as carriers, permeases, channels, and transporters.

A. Passive – Mediated Transport

- **Ionophores:** They are small organic molecules, many are antibiotics and vastly increase the permeability of membranes to particular ions. It can be carriers or channel formers. The efficiency of a carrier can be drastically reduced at lower T but not so for the channel formers.
- **Carriers:** The carrier-ionic complexes are soluble in non-polar solvents. It can move from one side to the other side of a lipid-bilayer membrane. Valinomycin contains 6 oxygen atoms complexing with a K^+. Channel Former: gramicidin two

molecules together form a channel. Facilitates the passage of cations (eg. Na^+, K^+, H^+) but is blocked by Ca^{2+}.

- **Bacterial Porins:** are channel forming proteins. X-ray structure is known (*E. coli* OmpF porin). Some are cation selective, others are anion selective.
- **Ion Channels:** It has many ion-specific channels for the rapid passage of ions. K^+ passively diffuses from cytoplasm to the extracellular space through transmembrane proteins "K^+ Channels." (High selectivity for K+ over Na^+ are "Gated." The physiological functions of ion channels depend on their ion specificity, speed of transport, and the ability to gate. Mechanosensitive channels: it responds to physical stimuli (touching and sound , etc.), which cause local deformations in membranes. Ligand-gated channels: respond to extracellular chemical stimuli such as neurotransmitters. Signal-gated channels: open upon intracellular binding of Ca^{2+} or other signal molecules. Voltage-gated channels: respond to changes in membrane potential, such as in the generation of nerve impulses.
- **Aquaporins:** Water passes through biological membranes extremely fast and aquaporins were discovered in 1992 by Peter Agre.
- **Transport Proteins:** It has functions in mediated passive and active transport.

20. CARBOHYDRATE METABOLISM: PHOTOSYNTHESIS: PHOTOSYNTHETIC PIGMENTS IN RELATION TO THEIR FUNCTIONS, PHOTOSYNTHETIC ELECTRON TRANSPORT AND GENERATION OF NADPH AND ATP, CYCLIC, NON-CYCLIC AND PSEUDOCYCLIC PHOTOPHOSPHORYLATION

Introduction

The flow of intermediates through metabolic pathways is controlled by four mechanisms

- Substrate availability
- Allosteric activation or inhibition
- Covalent modification of enzymes.
- Induction/repression of enzyme synthesis

The types of transporters and enzymes that exist in a certain cell type.

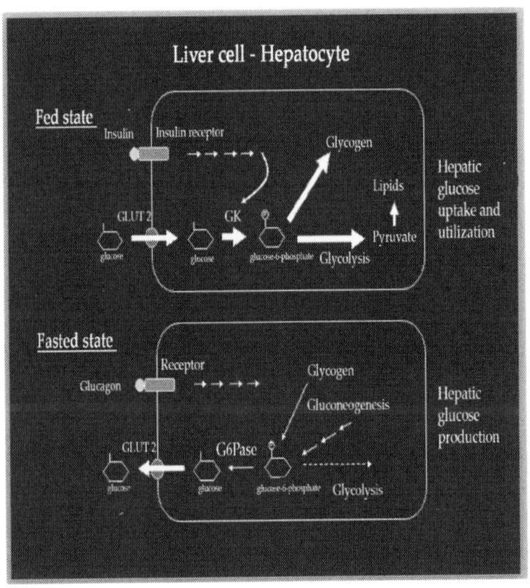

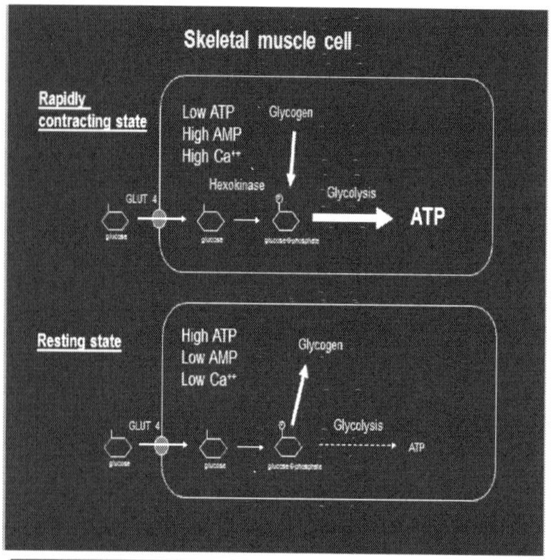

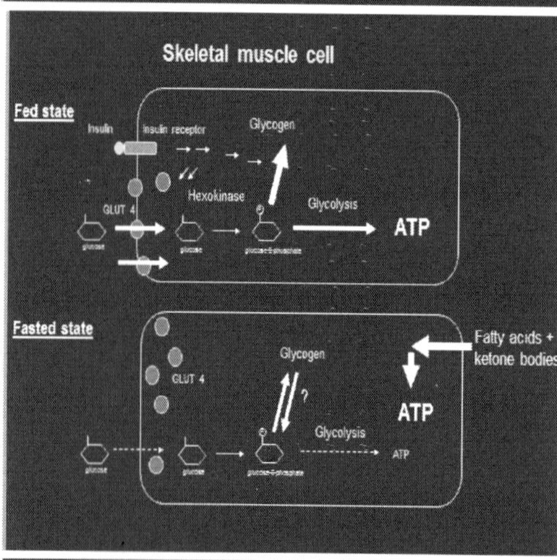

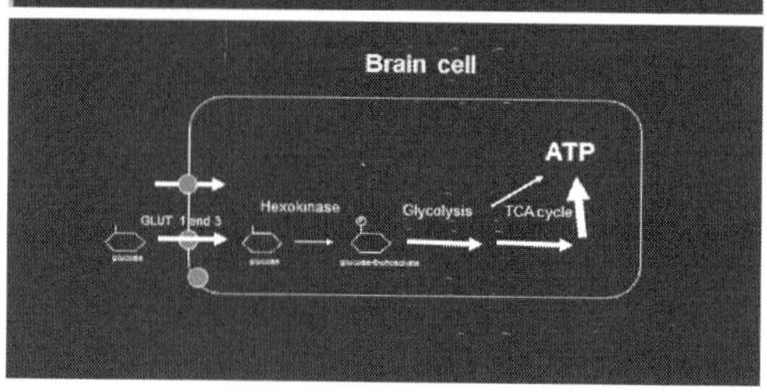

Concentration Effects on Enzymatic Steps

High K_m of glucokinase determines the rate of glucose catabolism and insulin secretion in pancreatic beta cells, High K_m of glucokinase allows liver cells to rapidly phosphorylate glucose at high concentrations.

Allosteric Effector

Glucose stimulates the release of glucokinase from the "glucokinase regulatory protein (GKRP) and glucose inhibits liver glycogen phosphorylase. Glucose stimulates gene transcription through, glucose-responsive regions of gene promoters. High glucose concentration stimulates the expression of the pyruvate kinase in the liver. The control of flux J through a pathway is distributed among all the enzymes. A key regulatory enzyme is characterized by a large change in flux when the enzymatic activity in the cell changes eg. phosphofructokinase in glycolysis. Once one enzymatic bottle neck for flux is removed, another enzyme may become limiting for the flux. Theoretically, if the enzyme activities of all the enzymes in the pathway were changed by the same degree, then the flux would change by the same degree. An efficient increase of flux through the pathway is brought about by increasing the enzyme activities or enzyme concentrations of all regulatory enzymes in a pathway. In the fed state with excess carbohydrates, the liver cells can synthesize fatty acids from glucose and fructose, and esterify them into triacylglycerols". Triacylglycerols are assembled together with cholesterol and protein into very-low density lipoprotein particles (VLDL) that are secreted into the circulation.

Conclusion

Glucose up-regulates genes encoding enzymes and proteins involved in the entire pathway of lipogenesis. ChREBP-Mlx is the principal mediator that transcriptionally regulates glucose-responsive genes.

21. CARBON REDUCTION IN C3 PLANTS – REACTIONS OF CALVIN CYCLE (REDUCTIVE PENTOSE PHOSPHATE PATHWAY), OXIDATIVE PENTOSE PHOSPHATE PATHWAY, REGULATION OF REDUCTIVE AND OXIDATIVE PENTOSE PHOSPHATE PATHWAY, PHOTORESPIRATION

Introduction

The net overall equation for photosynthesis is. Photosynthesis occurs in 2 "stages": The light reactions (or Light-Dependent Reactions). The Calvin cycle (or Calvin-Benson cycle or dark reactions or light-independent reactions) fig. 21.1

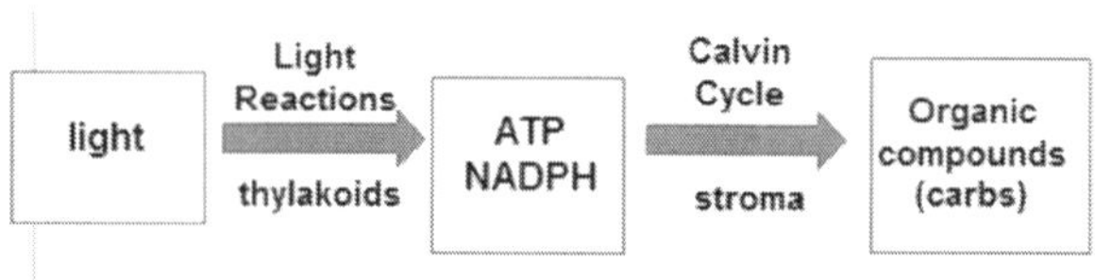

Figure 21.1. Photosynthesis.

In the Calvin cycle, chemical energy (from the light reactions) and CO_2 (from the atmosphere) are used to produce organic compounds (like glucose). The Calvin cycle occurs in the stroma of chloroplasts. The Calvin cycle involves the process of carbon fixation. This is the process of assimilating carbon from a non-organic compound (CO_2) and incorporating it into an organic compound.

Step 1: Carbon Fixation

There are three molecules of CO_2 (from the atmosphere) are joined to 3 molecules of RuBP (a 5-carbon sugar) by Rubisco (an enzyme also known as RuBP carboxylase).

Step 2: Reduction

The three 6-carbon molecules (very unstable) split in half, forming six 3-carbon molecules. These molecules are then reduced by gaining electrons from NADPH. ATP is required for this molecular rearranging There are now six 3-carbon molecules, which are known as G3P or PGAL. Since the

Calvin cycle started with 15 carbons (three 5-carbon molecules) and there are now 18 carbons, we have a net gain of 3 carbons. One of these "extra" 3-carbon G3P/PGAL molecules will exit the cycle and be used to form ½ a glucose molecule. Once the Calvin cycle "turns" twice (well, actually 6 times), those 2 molecules of G3P (3-carbon carbohydrate) will combine to form 1 molecule of glucose (6-carbon carbohydrate molecule) or another organic compound.

Step 3: Regeneration of RuBP

The remaining 5 G3P molecules (3-carbons each) get rearranged (using ATP) to form 3 RuBP molecules (5-carbons each).

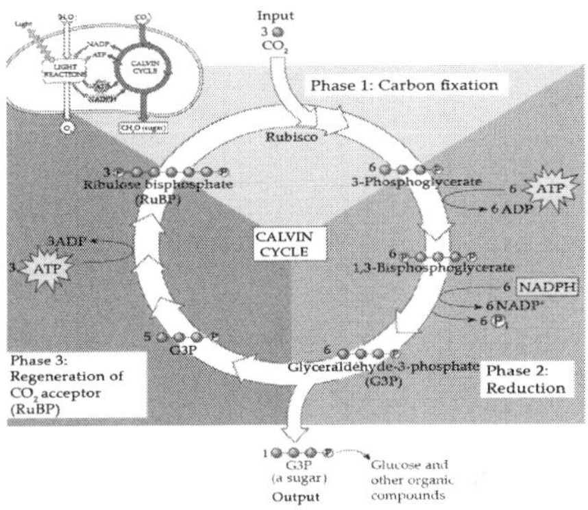

In the Calvin cycle, energy and electrons from the light reactions (in the form of ATP and NADPH) and carbon dioxide from the atmosphere are used to produce organic compounds. The Calvin cycle occurs in the stroma inside the chloroplasts (inside the cells). The carbon dioxide, ATP, and NADPH are required (reactants). The organic compounds (G3P) are produced. So, as a broad overview of photosynthesis. The Light reactions (Phase 1) capture the energy in sunlight and convert it to chemical energy in the form of ATP and NADPH through the use of photosystems, electron transport chains, and chemiosmosis. The Calvin cycle (Phase 2) uses the energy transformed by the light reactions along with carbon dioxide to produce organic compounds.

The Rate (Or Speed) Of Photosynthesis Can Vary, Based On Environmental Conditions

- Light intensity
- Temperature
- Oxygen concentration

Light intensity

As light intensity increases, so too does the rate of photosynthesis. This occurs due to the increased excitation of electrons in the photosystems. However, the photosystems will eventually become saturated. It is above this limiting level, no further increase in photosynthetic rate will occur.

Temperature

The effect of temperature on the rate of photosynthesis is linked to the action of enzymes. As the temperature increases up to a certain point, the rate of photosynthesis increases. The molecules are moving faster & colliding with enzymes more frequently and facilitating chemical reactions. However, at this temperature higher than this point, the rate of photosynthesis decreases.

Oxygen concentration

As the concentration of oxygen increases, the rate of photosynthesis decreases. This occurs due to the phenomenon of photorespiration.

Photorespiration

Photorespiration occurs when Rubisco (RuBP carboxylase) joins oxygen to RuBP in the first step of the Calvin cycle rather than carbon dioxide. Whichever compound (O_2 or CO_2) is present in higher concentration will be joined by Rubisco to RuBP. Photorespiration prevents the synthesis of glucose AND utilizes the plant's ATP. Photorespiration is primarily a problem for plants under water stress. When plants are under water stress, their stomata close to prevent water loss through transpiration. However, this also limits gas exchange. O_2 is still being produced (through the light reactions). Thus, the concentration of O_2 is increasing. CO_2 is not entering the leaf since the stomata are closed. Thus, as the CO_2 is being used up (in the Calvin Cycle) and not replenished, the concentration of CO_2 is decreasing. As the concentration of O_2 increases and the concentration of CO_2 decreases (due to the closure of the stomata to prevent excessive water loss), photorespiration is favored over photosynthesis. Some plant species that live in hot, dry climates (where photorespiration is an especially big problem) have developed mechanisms through natural selection to prevent photorespiration C_4 plants and CAM plants.

C_3 Plants

C_3 plants, which are "normal" plants, perform the light reactions and the "Calvin cycle in the mesophyll cells of the leaves. The bundle sheath cells of C_3 plants do not contain chloroplasts. C_4 plants and CAM plants modify the process of C_3 photosynthesis to prevent photorespiration. C_4 plants perform the Calvin cycle in a different location within the leaf than C_3 plants. CAM plants obtain CO_2 at a different time than C_3 plants. Both C_4 and CAM plants separate the initial fixing of CO_2 (carbon fixation) from the using of CO_2 in the Calvin cycle. Plants that use C_4 photosynthesis include corn, sugar cane, and sorghum. In this process, CO_2 is transferred from the mesophyll cells into the bundle-sheath cells, which are impermeable to CO_2. This increases the concentration of CO_2. Thus, the Calvin cycle is favored over photorespiration. The bundle-sheath cells of C_4 plants do contain chloroplasts. C_4 plants use the Hatch-Slack pathway prior to the Calvin cycle PEP carboxylase adds carbon dioxide to PEP, a 3-carbon compound, in the mesophyll cells. This produces a 4-carbon compound. This 4-carbon molecule then moves into the bundle-sheath cells via plasmodesmata. In the bundle sheath cells, the CO_2 is released and the Calvin cycle begins. Plants that use CAM photosynthesis include succulent plants (like cacti) and pineapples. In CAM (crassulacean acid metabolism) photosynthesis, plants open their stomata at night to obtain CO_2 and release O_2. This prevents them from drying out by keeping their stomata closed during the hottest & driest part of the day. When the stomata are opened at night, the CO_2 is converted to an organic acid (via the C_4 pathway) and stored overnight. During the day when light is present to drive the light reactions to power the Calvin cycle – carbon dioxide is released from the organic acid and used in the Calvin cycle to produce organic compounds. Even though the CO_2 is taken in at night, the Calvin cycle cannot occur because the light reactions can't occur in the dark. Both C_4 and CAM plants – which are primarily found in hot, dry climates have evolutionary adaptations which help prevent photorespiration. C_4 plants perform the Calvin cycle in the bundle sheath cells. CAM plants open their stomata at night and store the CO_2 until morning.

22. CARBON REDUCTION IN C4 PLANTS - LEAF ANATOMY OF C4 PLANTS, BIOCHEMICAL PATHWAY OF CO2 ASSIMILATION AND CAM PLANTS

Introduction

Photosynthesis is a process governing energy capture by ecosystems, and thus the entry of organic carbon into the biosphere controls on GPP: across ecosystems – LAI, N, growing season length. within ecosystems (daily and seasonal) – light, temp, nutrients, photosynthesis. Light harvesting converts light energy to chemical energy (ATP, NADPH) splits H_2O and produces O_2 as a byproduct. Carbon fixing uses ATP and NADPH from light reactions to fix CO_2 into sugar.

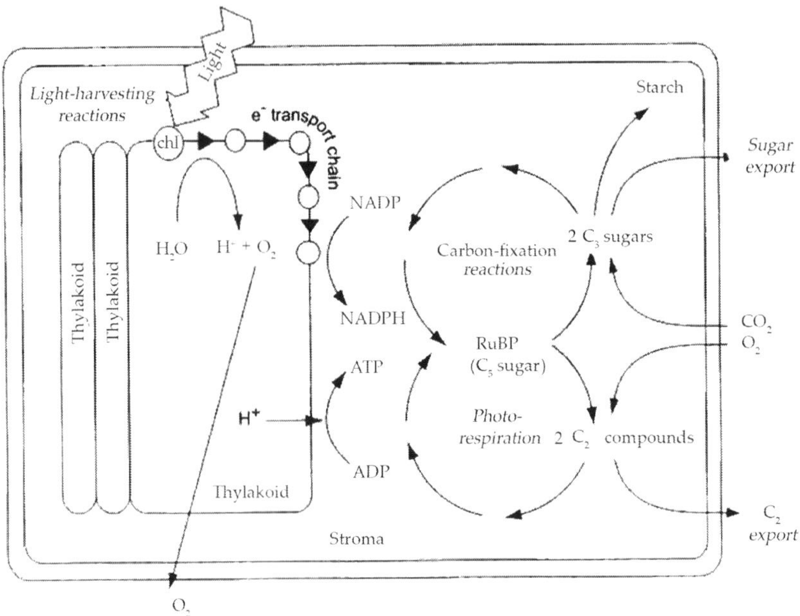

CO_2 enters leaf via diffusion

Stomata (plural), stoma (singular): biggest and most variable component of the diffusion barrier. Water evaporates on the cell surface, diffusing out through the stomata, leaf boundary layer, and into the bulk atmosphere is called transpiration.

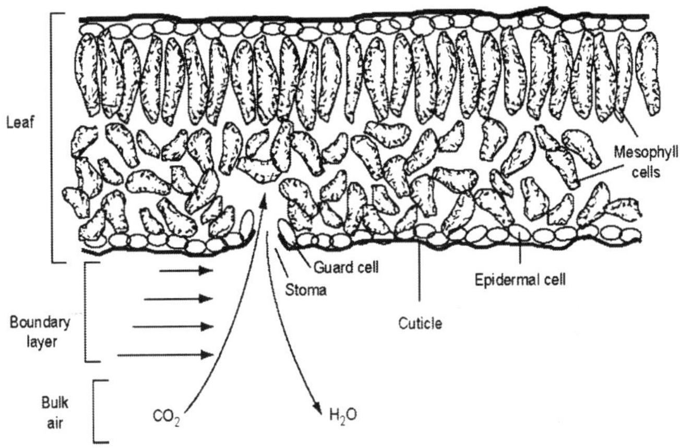

Stomata: carbon gain, water loss inevitable trade-off. Three major photosynthetic pathways. C3 photosynthesis 85% of plants. C4 photosynthesis: spatial separation of 2 carbon fixation paths. CAM (Crassulacean Acid Metabolism): The temporal separation of 2 carbon fixation paths.

3 Photosynthetic Pathways

C4 and CAM reduce water loss and reduce photorespiration because PEP carboxylase has a higher affinity for CO_2 and no affinity for O_2. These adaptations are most important in hot and dry environments.

The C4 and CAM pathways evolved to reduce the impact of oxidation by RUBISCO (photorespiration). The C4 photosynthetic pathway concentrates CO_2 in the vascular bundle sheath cells of leaves to significantly reduce competition with O_2 for the active site on RUBISCO. The CAM photosynthetic pathway further separates absorption of CO_2 in time from its incorporation in the Calvin cycle. C4 & CAM are more efficient than C3 at yielding energy above about 85 degrees F–Rubisco favors oxygenation the higher the temp because the concentration of CO_2 in leaf tissue is less the higher the temperature.

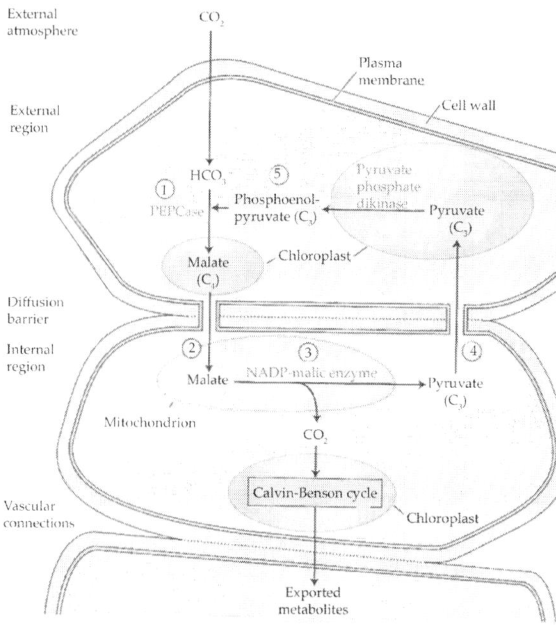

At 86° F C4 becomes more efficient than C3 due to photorespiration. The starch-rich chloroplasts of the Kranz anatomy lack grana, the site of the light reactions. They differ from the chloroplasts of the outer bundle sheath (dimorphic chloroplasts). In addition to being more efficient like C4 at higher temperatures, CAM conserves water by allowing stomata to be closed during the heat of the day. "Since every CO_2 molecule has to be fixed twice, first by 4-carbon organic acid and second by RuBisCO, the C4 pathway uses more energy than the C3 pathway. The C3 pathway requires 18 molecules of ATP for the synthesis of one molecule of glucose, whereas the C4 pathway requires 30 molecules of ATP. This energy debt is more than paid for by avoiding losing more than half of photosynthetic carbon in photorespiration as occurs in some tropical plants, making it an adaptive mechanism for minimizing the loss".

23. SUCROSE METABOLISM: TRANSPORT OF SUGAR FROM SOURCE TO SINK, PHLOEM LOADING AND UNLOADING OF SUCROSE, SUCROSE METABOLIZING ENZYMES, SUCROSE SYNTHESIS, SUCROSE STORAGE AND UTILIZATION, SUCROSE-STARCH INTER-CONVERSION

Introduction

It is a non-reducing disaccharide. It composed of glucose and fructose linked via their anomeric carbons. It is synthesized in the cytosol of plant cells. It is synthesized from UDP-glucose and fructose 6-phosphate. The sucrose is translocated from its site of synthesis in mature leaves to various metabolic tissues, where it is used to support the growth and synthesis of reserve materials such as starch fig. 23.1.

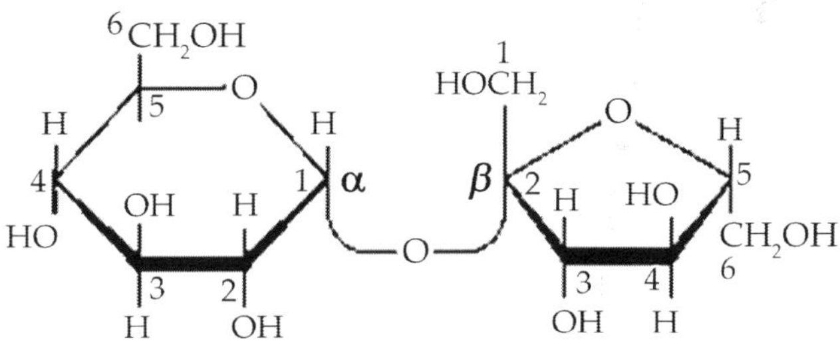

Sucrose

α-D- glucopyranosyl*β*-D- fructofuranoside

Glc(α 1 *β*)Fru

Figure 23.1. Sucrose metabolism.

Sucrose is synthesized from UDP-Glucose and Fructose-6-P in cytosol by sucrose 6-phosphate synthase and sucrose 6-phosphate phosphatase. "Sucrose biosynthesis is beginning with dihydroxyacetone phosphate exported from the chloroplast by Pi-triose phosphate antiporter. Dihydroxyacetone phosphate is then converted to glyceraldehyde 3-phosphate by triose phosphate isomerase. Pi-triose phosphate antiporter a transport system exports triose phosphates from the chloroplast and import phosphate: P_i-triose phosphate antiporter simultaneously moves P_i into the chloroplast and moves triphosphate into the cytosol. Sucrose synthesis releases P_i. If this exchange was blocked, triose phosphate synthesis would quickly deplete the available P_i in the chloroplast. A second potential source of energy is the ATP and NADPH generated in chloroplast". Pi-triose phosphate antiporter system has the indirect effect of moving ATP equivalents and reducing equivalents.

Trans-Aldolase Reaction

The condensation of dihydroxyacetone phosphate and glyceraldehyde -3-phosphate by transaldolases. Transaldolase reaction is identical to aldolase reaction in glycolysis/gluconeogenesis; other is unique to carbon assimilation. "Sucrose synthase (Km, 15 mM) has a much lower Km for sucrose compared with the neutral invertase (Km, 65 mM). Consequently, the sucrose synthase pathway may be relatively more important when sucrose availability is limiting. This pathway is also more energetically efficient, as the energy contained in the glycosidic linkage of the sucrose molecule is preserved. Thus, to metabolize one molecule of sucrose to the level of Triose-P requires the input of three ATP in the sucrose synthase pathway, compared with four ATP in the invertase pathway. Consequently, it may be beneficial to the cells to have the most efficient pathway operate when carbon supplies are limiting".

24. STARCH METABOLISM: STARCH BIOSYNTHESIS, REGULATION AND DEGRADATION

Introduction

It is a polymer of α-D-glucose. It occurs in two main forms: amylose, consisting of predominantly linear chains of glucose monomers linked by α1,4-glycosidic bonds, and amylopectin, in which the chains are branched by the addition of α1,6-glycosidic bonds. The fewer branches than glycogen. It is synthesized in the chloroplast (stroma). The precursor is activated ADP-glucose fig. 24.1.

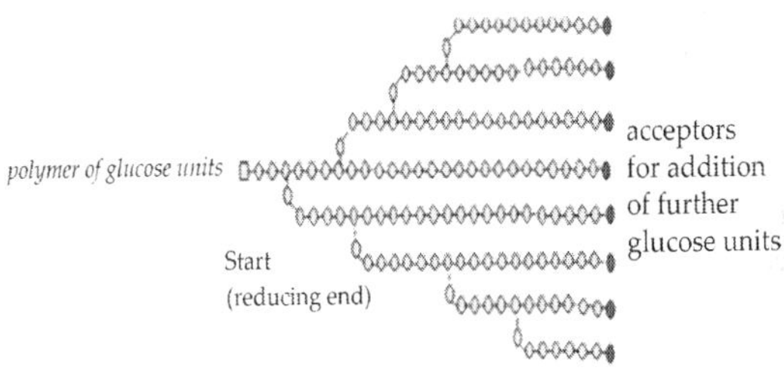

Figure 24.1. Polymer of α-D-glucose.

Starch granules are classified as transitory or reserve. The starch granules accumulate for only a short period of time before they are degraded eg. starch forms in leaf chloroplasts during the day. Hydrolyzed and transported to other plant parts at night as a simple sugar. The reserve starch, an energy storage for germination, a major component of food and feed, and an industrial starting material, is formed in amyloplasts. Starch is made in photosynthetic and non-photosynthetic cells fig 24.2.

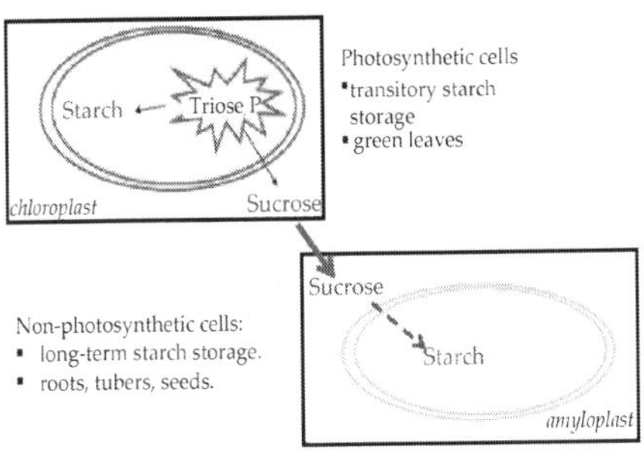

Figure 24.2. Photosynthetic and non-photosynthetic cells.

Amylose

A linear polymer of α-D-glucose with α1,4-linkage. It may have a low level of branching (~one branch per 1000 residues) with an α1,6-linkage. It comprises between 11 and 37% of the starch found in plants (depending upon the species and the site of storage). It is much lower in molecular weight than amylopectin.

Amylopectin

It is a highly branched polymer of α-D-glucose with α1,4& α1,6 linkages. It consists of 10,000 - 100,000 glucose units. It is highly branched, 20 - 25 glucose molecules/branch. It makes up ~65% of starch. It has a much higher molecular weight than amylose.

Starch Biosynthesis

ADP-glucose is used as the precursor. The starch synthase transfers the glucose unit to the non-reducing end of a preexisting primer. The branches in amylopectin are synthesized using branching enzymes. The synthesis of ADP-Glucose, catalyzed by ADP-glucose pyrophosphorylase, is rate limiting. Starch Synthase (SS) catalyzes a 1,4- linkage between the non-reducing end of glucan chain & Glc from ADP-Glc. SS can use both amylose and amylopectin as acceptors. The priming event not known: some evidence for protein primer, some evidence for de-novo synthesis.

Starch Degradation

The starch granule is attacked by the endoamylase α-amylase, which releases soluble linear and branched glucans. These are acted on by the debranching enzyme limit dextrinase and the exoamylase β-amylase to produce maltose. Maltose is then hydrolyzed to glucose by an α-glucosidase (maltase).

25. METABOLISM OF NITROGEN-AMINO ACIDS AND NUCLEOTIDES: PATHWAY OF NITRATE ASSIMILATION, INCORPORATION OF NH3 INTO AMINO ACIDS AND PROTEINS, BIOSYNTHESIS OF AMINO ACIDS, SULFATE ASSIMILATION, BIOSYNTHESIS OF NUCLEOTIDES – DE NOVO AND SALVAGE PATHWAY

Introduction

Nucleic acids are essential for information transfer in cells. The information encoded in a DNA molecule is transcribed via the synthesis of an RNA molecule. The sequence of the RNA molecule is "read" and is translated into the sequence of amino acids in a protein. The compound contained C, N, O, and a high amount of P. It was an acid compound found in nuclei therefore named nucleic acid. In 1944, Oswald, Avery, MacLeod and McCarty demonstrated that DNA is the molecule that carries genetic information. 1953 Watson and Crick proposed the double helix model for the structure of DNA nucleic acids are long polymers of nucleotides.

Nucleotides and Nucleic Acids

The amino acids sequence and nucleotide sequence in RNA is specified by a nucleotide sequence in the cell's DNA. **Gene:** segment of a DNA molecule that contains the information required for the synthesis of a functional biological product whether a protein or RNA. The cell contains thousands of genes and DNA molecules. **Nucleic acids** are polymers of nucleotides. **Deoxyribonucleic acid** - DNA - storage of genetic information. **Ribonucleic acid - RNA** - carriers of genetic information and catalysis. **Nucleotide structure**: Nitrogenous base+ ribose phosphate fig. 25.1

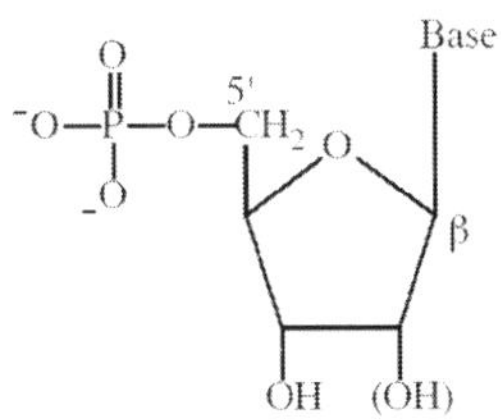

Figure 25.1. Nitrogenous base+ ribose phosphate.

Nucleotides Structure

The nucleotide has three characteristic components. **Nitrogenous base, Pentose sugar and Phosphate fig 25.2.**

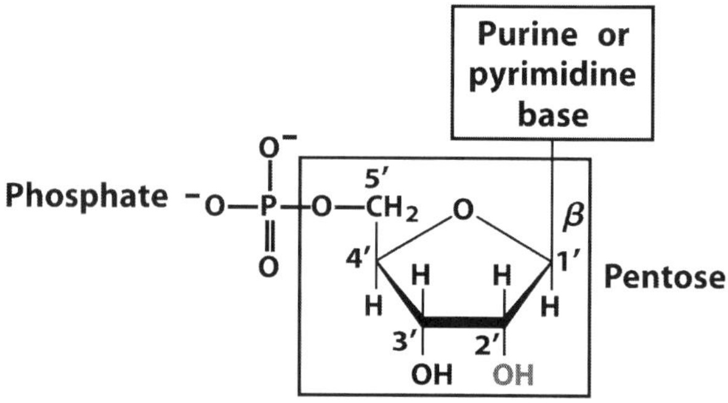

Figure 25.2. Nucleotides structure.

Nitrogenous Bases

Nitrogenous base: derivatives of purines and pyrimidines, DNA and RNA contain the same purine bases and the pyrimidine base cytosine but thymine found only in DNA and uracil found only in RNA fig 25.3.

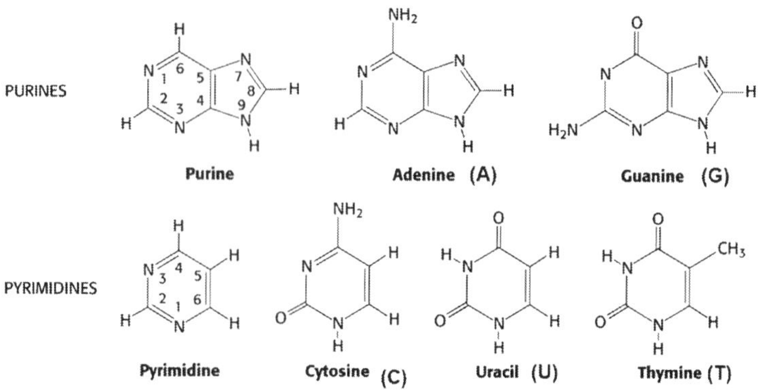

Figure 25.3. Nitrogenous base.

Pentoses of Nucleotides

D-ribose (in RNA), 2-deoxy-D-ribose (in DNA), The difference - 2'-OH vs 2'-H, this difference affects secondary structure and stability. The bases are attached by b-N-glycosidic linkages to 1 carbon of pentosc sugar – (Nucleoside). The base is linked via a b-N-glycosidic bond. The carbon of the glycosidic bond is anomeric. The named by adding -idine to the root name of a pyrimidine or -osine to the root name of a purine sugars make nucleosides more water-soluble than free bases.

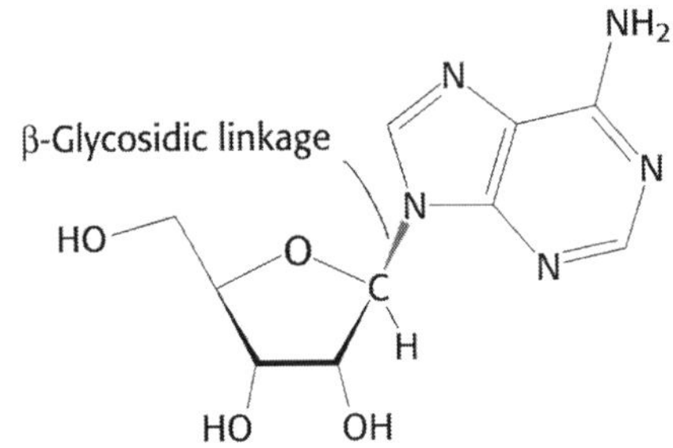

Nucleosides

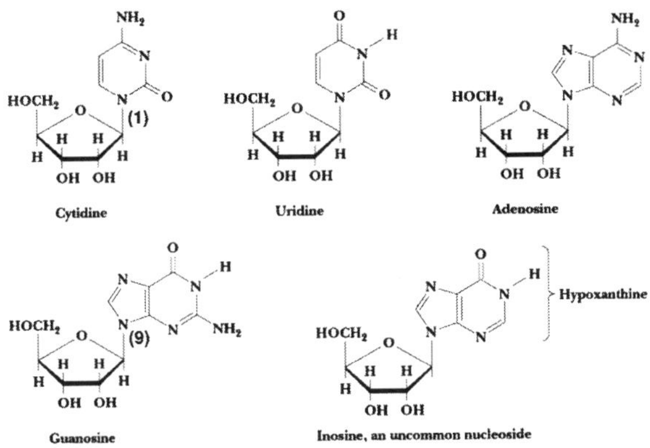

Ribonucleotides

- Adenosine 5'-monophosphate, **A**denylate, AMP
- Guanosine 5'-monophosphate, **G**uanylate, GMP
- Cytidine 5'-monophosphate, **C**ytidylate, CMP
- Uridine 5'-monophosphate, **U**ridylate, UMP

Deoxyribonucleotides

- Deoxythymidine 5' monophosphate
- Deoxythymidine, dTMP.
- NMP = Nucleoside mono phosphate.
- Numbering of Ribose sugar is given 1', 2', …, 5'

Unusual nucleotides

The modified nucleotides are found in some viral DNA and in transfer RNA. These modifications include methylation, hydroxymethylation, glycosylation, acetylation.

Functions of Nucleotides

Nucleoside 5'-triphosphates are carriers of energy, bases serve as recognition units, cyclic nucleotides are signal molecules and regulators of cellular metabolism and reproduction. They are the structural component of some coenzymes, e.g CoA, FAD, NADH, NADPH, ATP is central to energy metabolism, GTP drives protein synthesis, CTP drives lipid synthesis, UTP drives carbohydrate metabolism.

De Novo Synthesis of Purine Nucleotides

The atoms of the purine ring are contributed by a number of compounds including amino acids (aspartic acid, glycine and glutamine) CO2, and tetrahydrofolate. These compounds donate N and C to constructed Ribose 5-phosphate.

Synthesis of 5-phosphoribosyl-1-pyrophosphate (PRPP)

Ribose 5-phosphate is synthesized from HMP (Hexose monophosphate pathway), ribonucleotides are first synthesized then may be reduced to deoxyribonucleotides.

Synthesis of 5-phosphoribosyl-1-pyrophosphate (PRPP)

The amide group of the glutamine replaces the pyrophosphate group attached to PRPP, this reaction is mediated by Glutamine: phosphoribosyl pyrophosphate amidotransferase. This enzyme is inhibited by the end product of this pathway purine 5'-nucleotides AMP, GMP and IMP. This reaction is the committed step in purine nucleotide synthesis.

Synthesis Inosine Monophosphate (IMP)

IMP is the parent purine nucleotide. the synthesis of IMP requires 4 ATP molecules.

Inhibitors of Purine Synthesis

Specific inhibitors that inhibit the growth of rapidly growing microorganisms eg. Sulfonamides. Structural analogs for folic acid (methotrexate).

Conversion Nucleoside Monophosphate (NMP) to Nucleoside Diphosphate (NDP) and Triphosphate (NTP)

NDP and NTP are synthesized from the corresponding NMP by Nucleoside monophosphate kinases. These kinases don't discriminate between ribose or deoxyribose in the substrate. ATP is the source of the transferred phosphate, examples:

AMP + ATP ↔ 2 ADP ➔ adenylate kinase (highly active in the liver)

GMP + ATP ↔ GDP + ADP ➔ Guanylate kinase

Nucleoside diphosphates and triphosphates are interconverted by nucleoside diphosphate kinase

GDP + ATP ↔ GTP + ADP

CDP + ATP ↔ CTP + ADP

Salvage Pathway for Purines

Salvage Pathway

Purines that result from the normal turnover of cellular nucleic acids or that obtained from the diet and not degraded can be reconverted into nucleoside triphosphates and used by the body. The two enzymes are involved such as Adenine phosphoribosyl transferase (APRT) and hypoxanthine-guanine phosphoribosyl transferase (HGPRT). Both enzymes utilize PRPP as the source of the ribose-5-phosphate group.

Degradation of Purine Nucleotides

Purines are sequentially degraded into uric acids (in humans). Several steps will be involved in this catabolic pathway.

Degradation of Dietary Nucleic Acids in the Small Intestine

Ribonucleases and deoxyriboncleases secreted in the pancreatic juice can hydrolyze RNA and DNA into oligonucleotides. Oligonucleotides are further hydrolyzed by pancreatic phophodiesterases producing a mixture of 3'-and

5' mononucleotides. A family of nucleotidases removes the phosphate group releasing nucleosides that may absorbed from GIT. Dietary nucleotides are not used to large extent in cells because they are converted into uric acid in the small intestine and are also used by the normal flora.

Pyrimidine Nucleotides Synthesis

Purine ring is synthesized on an existing ribose 5-phosphate pyrimidine ring is synthesized then attached to ribose 5-phosphate donated by PRPP the sources of carbon atoms in pyrimidine rings are glutamine, CO_2, and aspartic acid.

Synthesis of Carbamoyl Phosphate

The committed step of this pathway in mammalian cells is the synthesis of carbamoyl phosphate from Glutamine and CO_2

$$2ATP + CO2 + Glutamine \leftrightarrow Carbamoyl\ phosphate + 2ADP + Glutamate$$

This reaction is mediated by Carbamoyl Phosphate Synthetase II (CPS II). CPS II is inhibited by UTP and activated by ATP and PRPP. Carbamoyl phosphate is the precursor of Urea and the pyrimidine synthesis occurs in the cytosol while the urea production occurs in the mitochondria by Carbamoyl Phosphate Synthetase I (CPS I), CPS I uses ammonia as a source of nitrogen, CPS II uses the amide group of glutamine, glutamine is required in the synthesis of both purines and pyrimidine.

Degradation of Pyrimidines

Purines are not cleaved in human cells. Pyrimidines rings can be opened and degraded to highly soluble structures, such as β-alanine and β-aminoisobutyrate that can serve as precursors of acetyl CoA and succinyl CoA. Pyrimidine can be salvaged and converted into nucleotides by the enzyme pyrimidine phosphoribosyltransferase and it utilizes the PRPP.

Conversion of Ribonucleotides to Deoxyribonucleotides

2'-deoxyribonucleotides are synthesized from ribonucleoside diphosphatase. Ribonucleotide reductase is multi-subunit enzyme (2B1 and 2B2 subunits) that catalyzes the reduction of NDP (ADP, GDP, CDP, UDP) into dNDP (dADP, dGDP, dCDP, dUDP). The immediate donors of hydrogen atoms needed for the reduction are two–SH groups of the enzyme itself. The reduced form of the enzyme should be regenerated. The reducing agent is a peptide coenzyme of ribonucleotide reductase called Thioredoxin. The thioredoxin contains two cysteine residues that can be oxidized to reduce the ribonucleotide reductase enzyme. The oxidized thioredoxin is reduced back by NADPH, this reaction is mediated by Thioredoxine reductase

Regulation of Deoxyribonucleotides Synthesis

The regulation of this enzyme is complex. The activity is regulated but also substrate specificity. The binding of dATP to an allosteric site called activity site inhibits the enzyme while the binding of ATP to this site activates the enzyme. The binding of NTP to an allosteric site called substrate specificity will increase the conversion of different NTP to their corresponding dNTP according to the need of the cell. When ATP or dATP are bound to this site for the reduction of UDP, and CDP is favored. When dTTP, or dGTP is bound for the reduction of GDP, ADP is stimulated. Synthesis of Thymidine monophosphate from dUMP. dUMP is converted into dTMP by thymidylate synthetase which utilizes N^5 N^{10}–methylene tetrahydrofolate as the source of methyl group and 2 hydrogens is oxidation into dihydrofolate. inhibitors of thymidylate synthetase (5-floro uracil) act as anti-tumor. DHF can be reduced into THF by DHF reductase, which can be inhibited by methotrexate to inhibits purine synthesis and decrease the supply of THF so prevents methylation of dUMP to dTMP.

Nucleotide Biosynthesis

Nucleotides can be synthesized de novo ("from the beginning") from amino acids, ribose-5-phosphate, CO_2, and NH_3. Nucleotides can be salvaged from RNA, DNA, and cofactor degradation. Many parasites (e.g., malaria) lack de novo biosynthesis pathways and rely exclusively on salvage. The compounds that inhibit salvage pathways are promising antiparasite drugs.

de-Novo Biosynthesis of Nucleotides

All organisms were studied about bases synthesized while attached to ribose approximately the same. Glu provides most amino groups. Gly is the precursor for purines. Asp is precursor for pyrimidines. Nucleotide pools are kept low, so cells must continually synthesize them. This synthesis may actually limit rates of transcription and replication. Adenine and guanine are synthesized as AMP and GMP. Synthesis begins with reaction of 5-phosphoribosyl 1-pyrophosphate (PRPP) with Glu. Purine ring builds up following the addition of three carbons from glycine. The first intermediate with a full purine ring is inosinate (IMP). de-Novo synthesis of pyrimidine nucleotides (1, 2 and 3). Unlike purine synthesis, pyrimidine synthesis proceeds by first making the pyrimidine ring and then attaching it to ribose 5-phosphate. Aspartate and carbamoyl phosphate provide the atoms for the ring structure. After the addition of ribose-5-phosphate via PRPP, the resulting nucleotide (orotidylate) is decarboxylated to form uridylate (UMP), the first possible pyrimidine. UMP is phosphorylated to UTP. After the formation of UTP, amination can convert UTP to CTP.

Ribonucleotides are Precursors to Deoxyribonucleotides

2'C-OH bond is directly reduced to a 2'-H bond without activating the carbon. catalyzed by ribonucleotide reductase. Two H atoms are donated by NADPH and carried by proteins thioredoxin or glutaredoxin.

Reduction of Ribonucleotides by Ribonucleotide Reductase

NADPH serves as the electron donor. It is directed through glutathione (a) or thioredoxin pathways (b).

dTMP is made from dUTP

- dUTP is made.
- dUTP → to dUMP by dUTPase
- dUMP → dTMP by thymidylate synthase

  - adds a methyl group from tetrahydrofolate

Thymidylate synthase is a target for some anticancer drugs.

Folic Acid Deficiency Leads to Reduced Thymidylate Synthesis

Folic acid deficiency is widespread, especially in nutritionally poor populations. It reduced thymidylate synthesis causes uracil to be incorporated into DNA. The repair mechanisms remove the uracil by creating strand breaks that affect the structure and function of DNA. It is associated with cancer, heart disease and neurological impairment.

Antibiotics also Target Nucleotide Biosynthesis

Allopurinol, and so on studied against African sleeping sickness (trypanosomiasis) because the trypanosomes lack enzymes for de novo nucleotide synthesis. Trimethoprim inhibits bacterial dihydrofolate reductase but binds the human enzyme several orders of magnitude less strongly

26. PROTEIN SYNTHESIS, MAJOR COMPONENTS OF PROTEIN SYNTHESIZING MACHINERY, STEPS IN POLYPEPTIDE SYNTHESIS.

Introduction

Proteins are important, 44% of the dry weight of the human body and 5% of human caloric intake goes for protein synthesis. It is catalyzing most of the reactions in living organisms. It serves many roles (enzymatic, structural, transport, regulation) protein synthesis is tightly regulated by environmental stimuli as well as intrinsic processes (e.g., hormonal, developmental). The dysregulation can cause disease and many antibiotics act at the level of protein synthesis.

Central Dogma

The central dogma states that once "information" has passed into protein it cannot get out again. The transfer of information from nucleic acid to nucleic acid, or from nucleic acid to protein, may be possible, but transfer from protein to protein, or from protein to nucleic acid is impossible. Information means here the precise determination of sequence, either of bases in the nucleic acid or of amino acid residues in the protein.

Genetic Code

The first step in decoding genetic messages is "**transcription**, during which a nucleotide sequence is copied from DNA to RNA. The next step is to join amino acids together to form a protein. The order in which amino acids are joined together determine the shape, properties, and function of a protein. The four bases of RNA form a language with just four nucleotide bases: adenine (A), cytosine (C), guanine (G), and uracil (U). The genetic code is read in three-base words called **codons**. Each codon corresponds to a single amino acid (or signals the starting and stopping points of a sequence). In **transcription**, a DNA sequence is rewritten, or transcribed, into a similar RNA. In eukaryotes, the RNA molecule must undergo processing to become a mature messenger RNA (mRNA). In **translation**, the sequence of the mRNA is decoded to specify the amino acid sequence of a polypeptide". The name translation reflects that the nucleotide sequence of the mRNA sequence must be translated into the completely different "language" of amino acids.

27. LIPID METABOLISM, SATURATED AND UNSATURATED FATTY ACID BIOSYNTHESIS, BIOSYNTHESIS OF TRIACYLGLYCEROLS AND PHOSPHOLIPIDS, CATABOLISM OF LIPIDS – LIPID DEGRADING ENZYMES, FATTY ACID OXIDATION

Introduction

The substances in dairy foods, known as fats and oils, hydrophobic and soluble in ethanol like alcohol, petroleum and chloroform, it is classified into two general types. Fats and waxes (can be hydrolyzed with ester linkages) cholesterol and steroids (cannot be hydrolyzed).

Type of Lipids

The saturated fat, animal oil like meat, milk, butter, vegetable oil like coconut and palm kernel oil, Polyunsaturated fat, plant sources like safflower, corn, cottonseed, sunflower oil and soybean oil Monounsaturated fat, plant and animal product like olive oil, canola oil, avocado and peanut oil.

Lipids Functions

- Excellent energy reserves
- Structure of cell membranes
- Organ padding
- Body thermal insulation
- Essential fatty acids (EFA)
- Hormone synthesis
- Fat soluble vitamin absorption

Lipids Disorder

- Lipids deficiency (Shortage in Lipids intake)
- Lipids exceeding (Overtaking in Lipids intake)

Lipids Deficiency

- Fat should comprise of 3% of total calories to prevent fatty acid deficiency.

- Fatty acid deficiency syndromes.
- Dry scaly skin, dermatitis (Linoleic acid deficiency).
- Hand tremors (Prostaglandin deficiency).
- Inability to control blood pressure.

Lipids Exceeding

Fat should comprise not more than 30% of total calories to prevent lipids exceeding, to prevent overtaking, we should consume fat breakdown (% total calories) <8% from saturated fat. It is 10% from polyunsaturated fat, 10-15% from monounsaturated fat. Fatty acids are taken up by cells. They may serve as: precursors in the synthesis of other compounds fuels for energy production substrates for ketone body synthesis. Ketone bodies may be exported to other tissues are used for energy production. Some cells synthesize fatty acids for storage or export. Fats are an important source of calories. Typically, 30-40% of calories in the American diet are from fat. Fat is the major form of energy storage. Typical body fuel reserves are: fat: 100,000 kcal. protein: 25,000 kcal. carbohydrate: 650 kcal. It provides 60% of the energy needs for the body at rest. TAG reserves would enable someone to survive starvation for ~30 days.

Digestion and absorption of lipids, 98% of ingested lipids are triacylglycerols (TAGs), Digestion in the mouth: enzymes are aqueous -little effect on lipids. Digestion in the stomach causes a large physical change: churned into droplets: "Chyme". Gastric Lipase begins actual lipid digestion. ~10% of TAGs are hydrolyzed in the stomach. Chyme stimulates cholecystokinin (CCK) to release bile from the gallbladder. Bile is an emulsifier

Fatty acid micelle hydrophobic fatty acids & monoacylglycerols are in the interior. Bile salts on the exterior. Micelles are small enough to penetrate the membrane of intestinal cells. Free fatty acids & monoacylglycerols are reformed into triacylglycerols. TAGs are combined with membrane & water-soluble proteins to form a chylomicron, a lipoprotein. Chylomicrons carry TAGs from intestinal cells into the bloodstream via the lymph system. Triacylglycerols reach the bloodstream & are hydrolyzed down to glycerol and fatty acids. These are absorbed by cells and processed further for energy by forming acetyl CoA or stored as lipids in fat cells (adipose tissue).

Triglyceride Storage & Mobilization

Storage of triacylglycerol is in adipocytes. Fatty acids are stored primarily as triacylglycerol. Triacylglycerol is hydrolyzed to release fatty acids when needed. Adipocytes are found mostly in the abdominal cavity and subcutaneous tissue. Store energy, insulation against heat loss, shock absorber for organs. Adipocytes are metabolically very active: triacylglycerol is constantly hydrolyzed & re-synthesized.

Hormonal control of lipolysis is the breakdown of triglycerides by lipases is under hormonal control. Hormones involved such as Epinephrine, glucagon, and insulin. Epinephrine & glucagon are promoting the breakdown of fat (lipolysis) Insulin inhibits lipolysis. Triacylglycerol mobilization is hydrolyzing lipid reserves in adipose tissue for energy triggered by hormones. ~10% TAGs are replaced in adipose tissue daily as they get used up for energy. Hydrolysis of stored triacylglycerols in adipose tissue is triggered by hormones that stimulate cAMP production within adipose cells. The third time is a charm. TAGs hydrolyzed a 3rd time to form fatty acids. Triacylglycerol lipase, Diacyclglycerol lipase, Monoacylglycerol lipase, only triacylglycerol lipase is activated by epinephrine.

28. CATABOLISM, GLYCOLYSIS, GLUCONEOGENESIS, CITRIC ACID CYCLE, ELECTRON TRANSPORT CHAIN AND OXIDATIVE PHOSPHORYLATION, INTERRELATIONSHIPS OF METABOLIC PATHWAYS, BIOCHEMISTRY OF SEED GERMINATION AND DEVELOPMENT AND BIOCHEMISTRY OF FRUIT RIPENING

Introduction

Metabolism is considered into metabolic pathways that break-down the molecules that generate energy (catabolism) and the molecules that consume energy by structure larger molecules (anabolism). Energy metabolism mentions additional exactly to the metabolic pathways that release or store energy fig. 28.1.

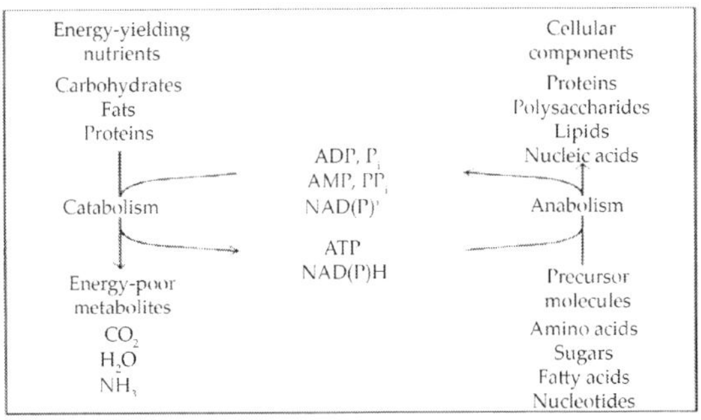

Figure 28.1. Metabolism.

The exergonic reactions of catabolism sustain ATP, NADH and NADPH formation. ATP and NADPH provide energy and reducing equivalents for the endergonic reactions of anabolism.

Fate of Glucose

The fate of glucose differs with physiological conditions, tissues, and organisms.

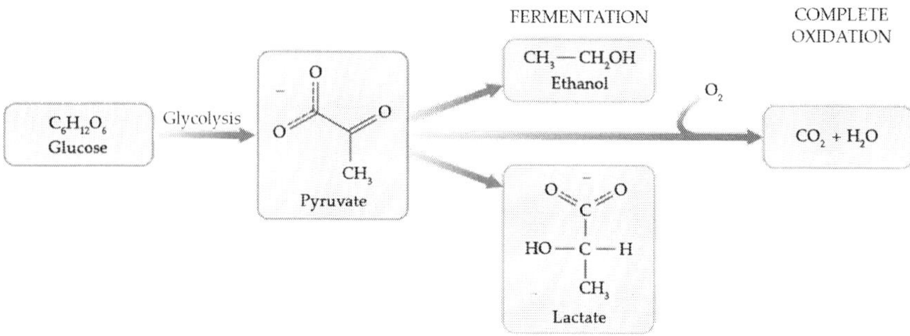

Glycolysis

Stage 1: Investment of ATP. Glucose is phosphorylated. The negative charge concentrates glucose in the cell and glucose becomes less stable.

Stage 2: The 6 carbon sugar is split into two 3-carbon fragments.

Stage 3: Energy yielding phase. The oxidation of the 3-carbon fragments yields ATP fig. 28.2 .

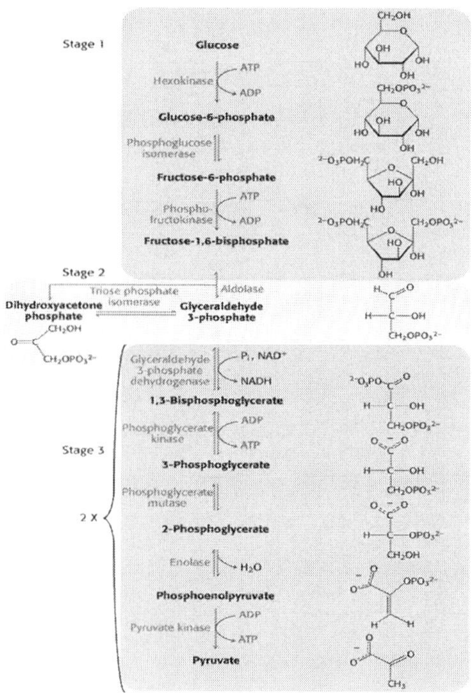

Figure 28.2. Glycolysis.

ΔG°'= -4.0 kcal mol-1

Phosphoryl transfer reaction: Kinases transfer phosphate from ATP to an acceptor. Hexokinase has a more general specificity in that it can transfer phosphate to other sugars such as mannose.

Phosphoglucose isomerase

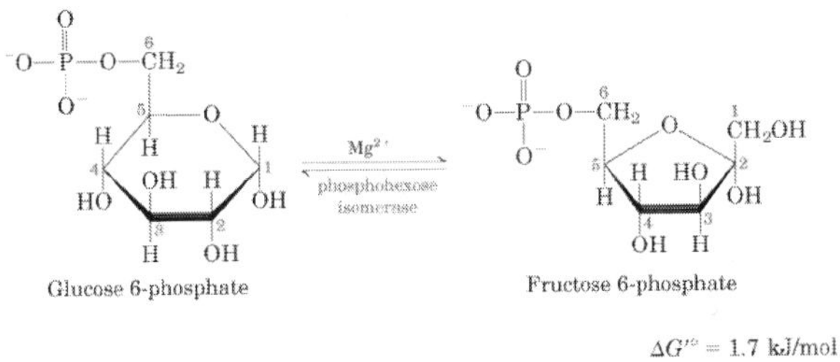

ΔG°'= .40 kcal mol-1

Phosphoglucose isomerase

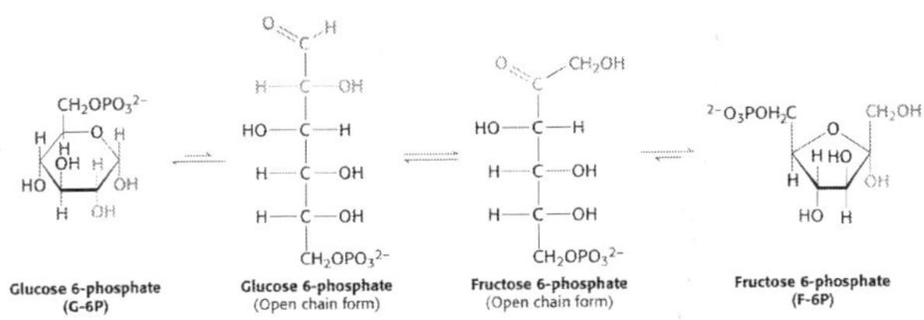

The enzyme opens the ring, catalyzes the isomerization, and promotes the closure of the five-member ring.

Phosphofructokinase-1

The 2nd investment of an ATP molecule in glycolysis.

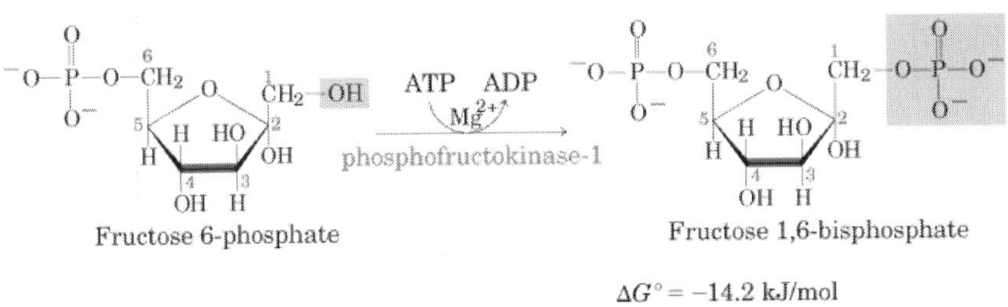

PFK is an important allosteric enzyme regulating the rate of glucose catabolism and plays a role in integrating metabolism. **Bis** means two phosphate groups on two different carbon atoms. Di means two phosphate groups linked together on the same carbon atom.

Phosphofructokinase-1

The PFK reaction is the first unique and irreversible step in the glycolytic pathway. It is the "committed" in the pathway. In general, the enzyme catalyzing the committed step in a metabolic pathway is the most important control component in the pathway.

2nd Stage: Aldolase

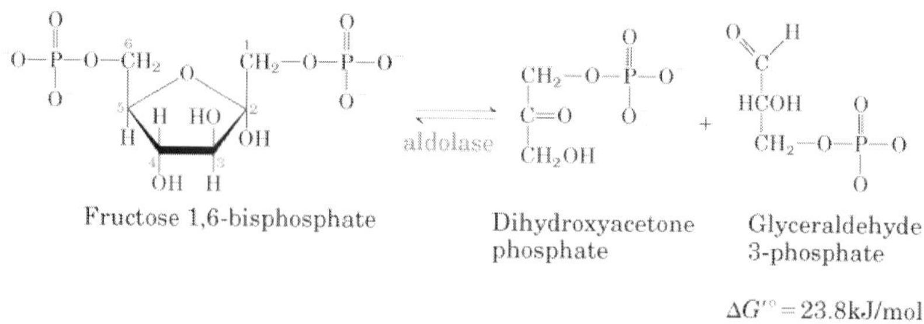

Reverse aldol condensation; converts a 6 carbon atom sugar to 2 molecules, each containing 3 carbon atoms.

Triose Phosphate Isomerase

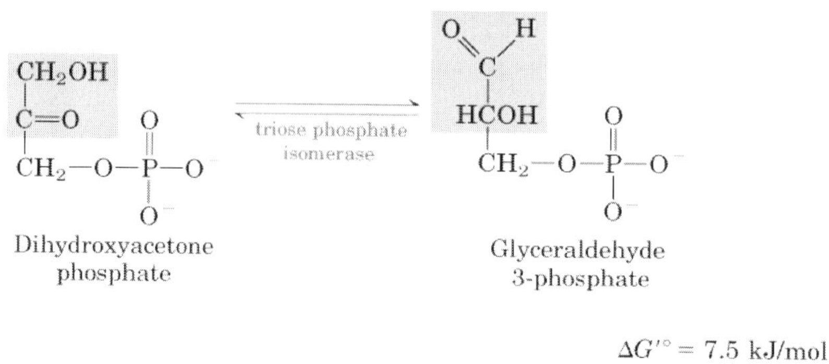

$\Delta G°' = 1.8$ kcal mol^{-1}

All the DHAP is converted to glyceraldehyde 3-phosphate. Although, the reaction is reversible it is shifted to the right since glyceraldehyde 3-phosphate is a substrate for the next reactions of glycolysis. Thus, both 3-carbon fragments are subsequently oxidized. The fate of glyceraldehyde 3-phosphate Glyceraldehyde 3-phosphate DH.

Stage 3: The energy-yielding phase

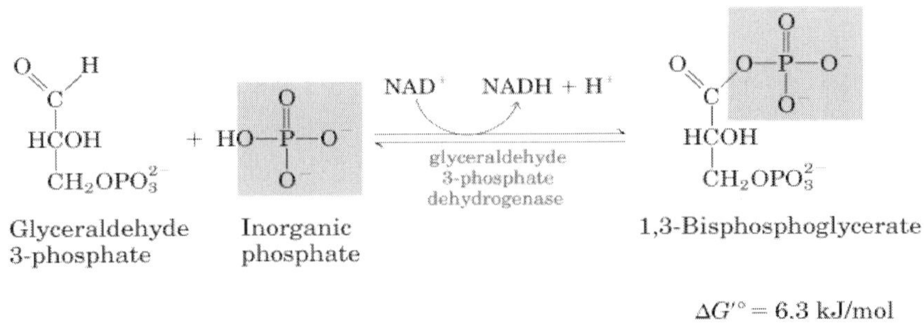

An aldehyde is oxidized to carboxylic acid and inorganic phosphate is transferred to form acyl-phosphate. NAD^+ is reduced to NADH.

ΔG°' = 1.5 kcal mol^{-1}

1,3-BPG has a high phosphoryl-transfer potential. It is a mixed anhydride. Notice, under anaerobic conditions NAD^+ must be re-supplied.

Phosphoglycerate Kinase Substrate-Level Phosphorylation

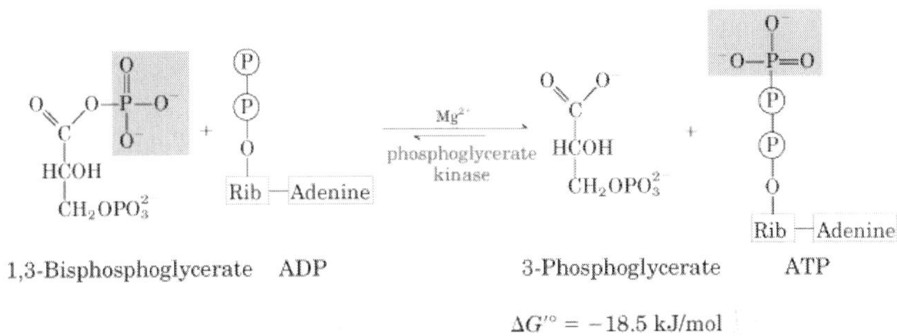

ΔG°' = -4.5 kcal mol^{-1}

ATP is produced from P_i and ADP at the expense of carbon oxidation from the glyceraldehyde 3-phosphate DH reaction. The two molecules of ATP are produced per glucose. At this point, 2ATPs were invested and 2ATPs are produced.

Phosphoglycerate Mutase

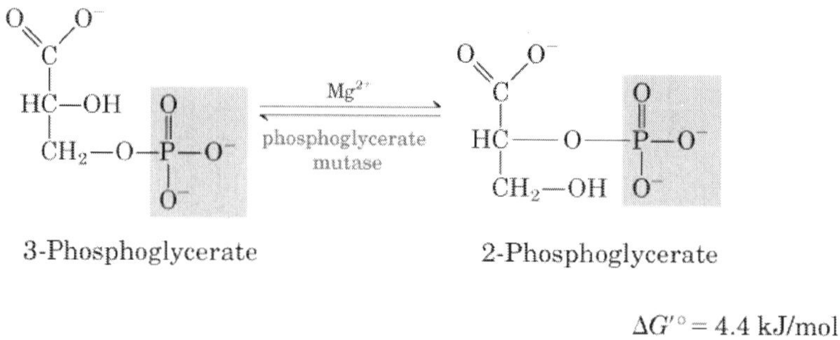

$\Delta G°' = 1.1$ **kcal mol^{-1}**

Enolase

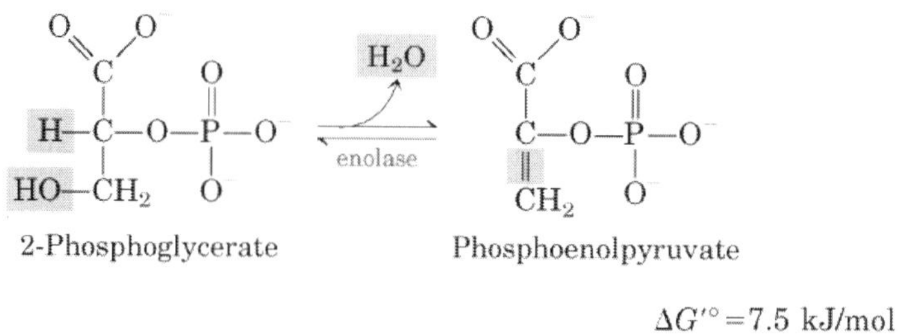

$\Delta G°' = .4$ **kcal mol^{-1}**

PEP

Pyruvate kinase 2nd example of substrate-level phosphorylation. The net yield from glycolysis is 2 ATP.

unstable Enol form→ more stable keto form

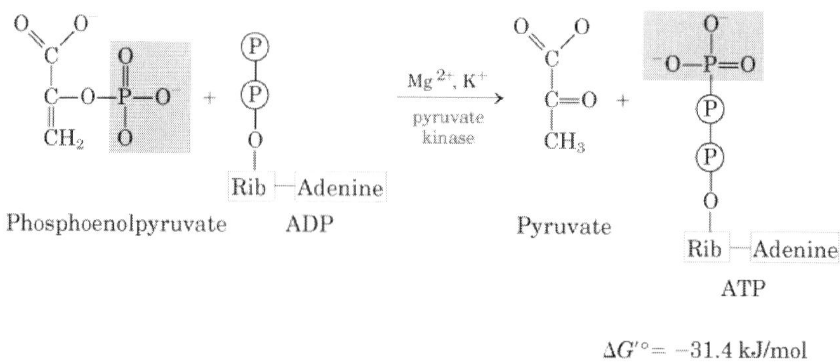

Substrate level phosphorylation is the synthesis of ATP from ADP that is not linked to the electron transport system. The conversion of glucose to pyruvate

Glucose + 2 P_i + 2 ADP + 2 NAD^+ → 2 pyruvate + 2 ATP + 2 NADH +2 H^+

The energy released from the anaerobic conversion of glucose to pyruvate is -47kcal mol^{-1}. Under aerobic conditions, much more chemical bond energy can be extracted from pyruvate. Under anaerobic conditions, pyruvate is converted to lactate. Exercising muscle is an example. The NAD^+ that is consumed in the glyceraldehyde 3-phosphate reaction is produced in the lactate DH reaction. The redox balance is maintained. The activities of glyceraldehyde 3-phosphate DH and lactate DH are linked metabolically.

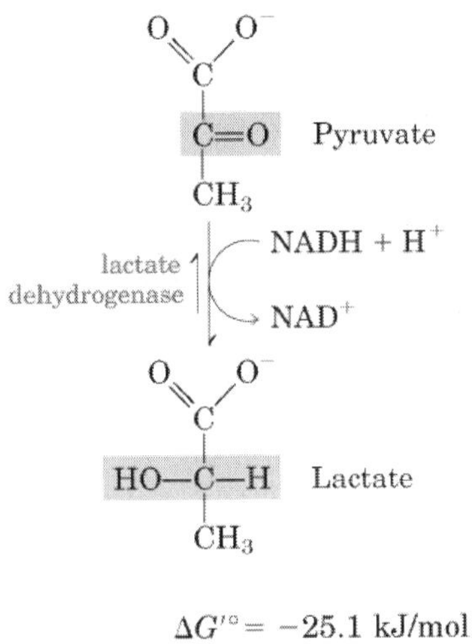

The NAD^+ that is consumed in the glyceraldehyde 3-phosphate reaction is produced in the lactate DH reaction. The redox balance is maintained. The activities of glyceraldehyde 3-phosphate DH and Lactate DH are linked metabolically. The NAD^+ that is consumed in the glyceraldehyde 3-phosphate reaction is produced in the lactate DH reaction. Thus, redox balance is maintained. The NADH that is produced in the glyceraldehyde 3-phosphate reaction is consumed in the lactate DH reaction. Thus, redox balance is maintained.

Glucose + 2 P_i +2 ADP → 2 lactate + 2 ATP + 2 H_2O

Fructose and galactose feed into the glycolytic pathway. In the liver, when fructose enters glycolysis the PFK reaction is bypassed.

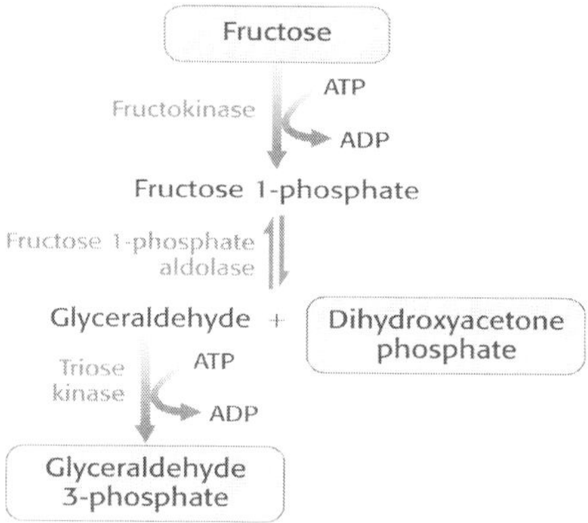

Galactose is converted to glucose-6P via a four-step reaction involving UDP-glucose.

In Summary

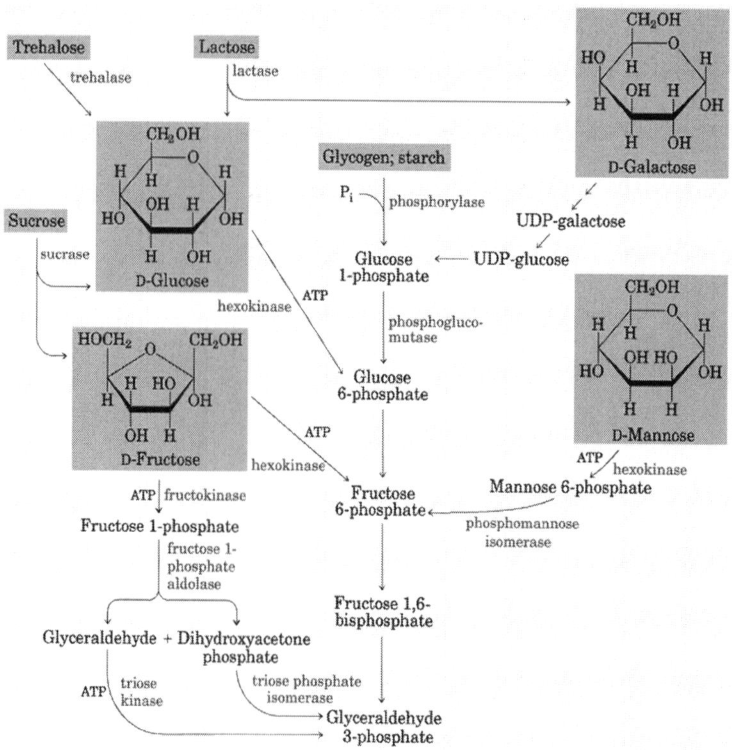

Gluconeogenesis

Gluconeogenesis is the synthesis of glucose from non-carbohydrate precursors. Glucose stores are depleted during periods of starvation or fasting beyond a day. Since the brain relies on glucose (120g/d) as a source of energy, glucose must be synthesized from molecules other than carbohydrates.

PYRUVATE → GLUCOSE

So any molecule that can be converted to pyruvate is considered glucogenic. Lactate and alanine are glucogenic. Glycerol is also glucogenic.

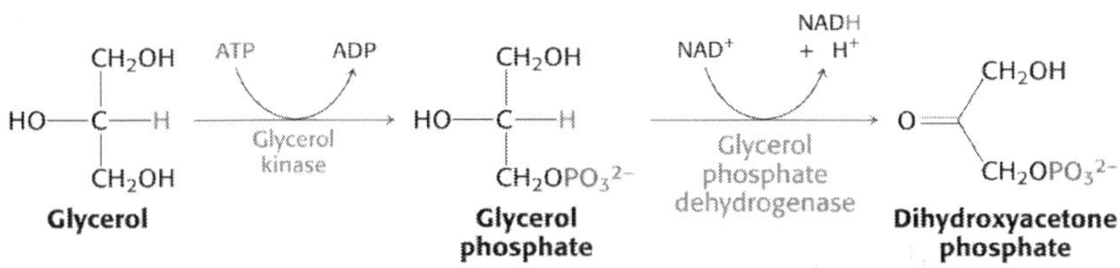

Gluconeogenesis: Pyruvate → Glucose

The enzymes in red belong to the gluconeogenic pathway. These reactions overcome the high negative free energy of the irreversible reactions of glycolysis. The enzymes in blue are held in common between the two pathways.

The irreversible glycolytic enzymes are: hexokinase (ΔG =-8 kcal mol^{-1}) phosphofructokinase (ΔG = -5.3 kcal mol^{-1}) pyruvate kinase (ΔG = -4.0 kcal mol^{-1}). The enzymes of gluconeogenesis are: pyruvate carboxylase (ATP) phosphoenolpyruvate carboxykinase (GTP) fructose 1,6-bisphosphatase, glucose 6-phosphatase

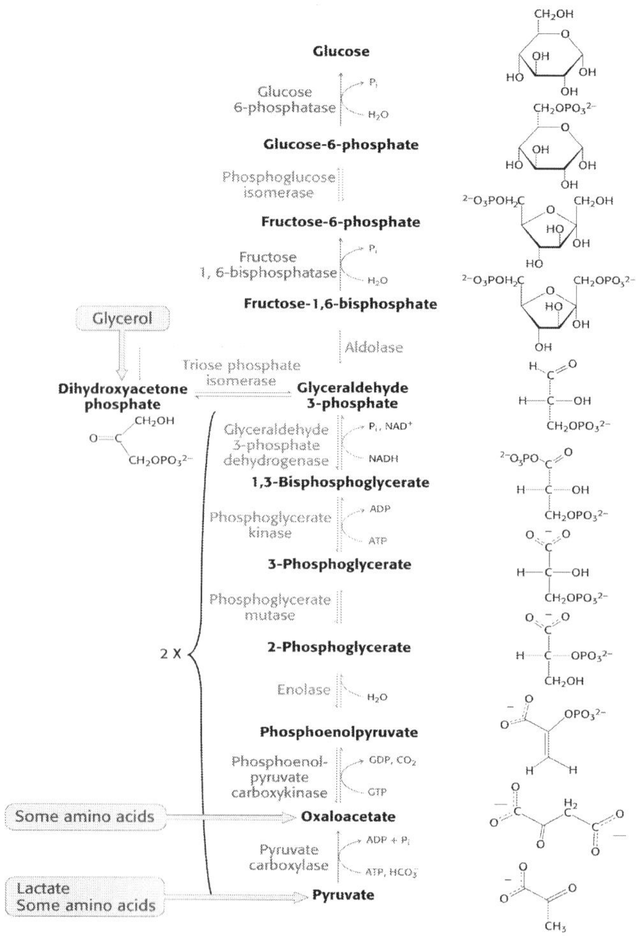

Pyruvate Carboxylase

Pyruvate + CO_2 + ATP + H_2O → oxaloacetate + ADP + P_i + 2 H^+.

Pyruvate Carboxylase fixes CO_2. Enzymes that fix CO_2 require the cofactor BIOTIN. Biotin is a vitamin and is always involved in CO_2 fixation. This reaction takes place in the mitochondrial matrix.

Biotin

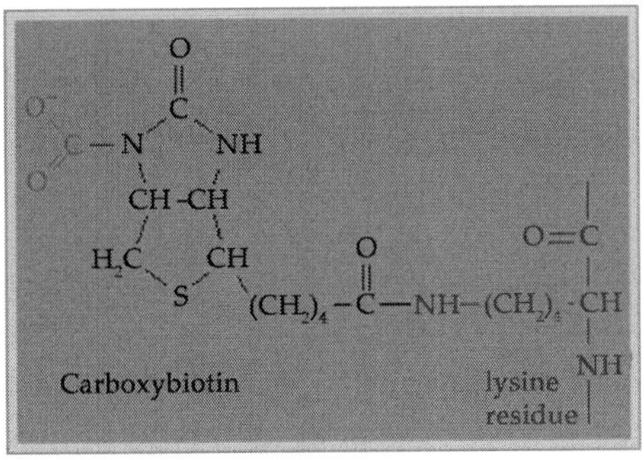

Notice there are no components of ATP in the structure of biotin. Phosphoenolpyruvate Carboxykinase

Oxaloacetate + GTP → phosphoenolpyruvate + GDP + CO_2

This reaction takes place in the cytosol. PEP is now synthesized and the sum of the two reaction is:

Pyruvate + ATP + GTP + H_2O →

PEP + ADP + GDP + P_i + H^+.

Pyruvate is carboxylated in the mitochondria. Pyruvate Carboxylase. Oxaloacetate can't pass out of the mitochondria. Oxaloacetate is decarboxylated and phosphorylated in the cytosol. Phosphoenolpyruvate Carboxykinase.

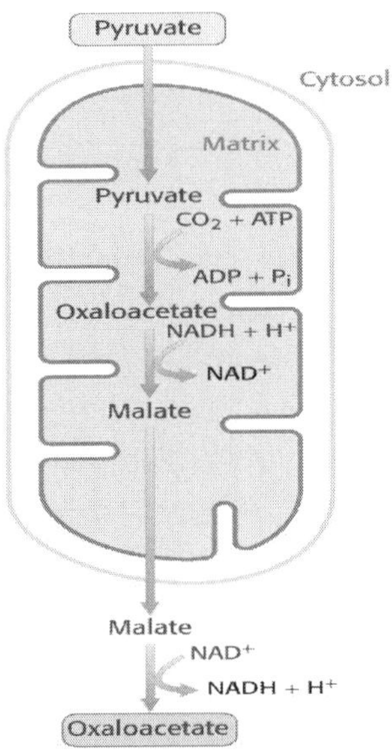

Fructose 1,6-Bisphosphatase

fructose 1,6-bisphosphate + $H_2O \rightarrow$ fructose 6-phosphate + P_i

Fructose 1,6-bisphosphatase is an allosteric enzyme and regulates gluconeogenesis. fructose 6-phosphate is easily converted to glucose 6-phosphate.

Glucose 6-Phosphatase

glucose 6-phosphate + $H_2O \rightarrow$ glucose + P_i.

The liver can send glucose to blood to maintain homeostasis. Glucose 6-phosphate is also a precursor to glycogen. Glucose 6-P is valuable; a precursor for glycogen synthesis. Glucose 6-phosphatase is present only in tissues responsible for maintaining blood glucose levels, liver and kidney. In the liver, glucose 6-phosphatase is highly regulated.

In Liver

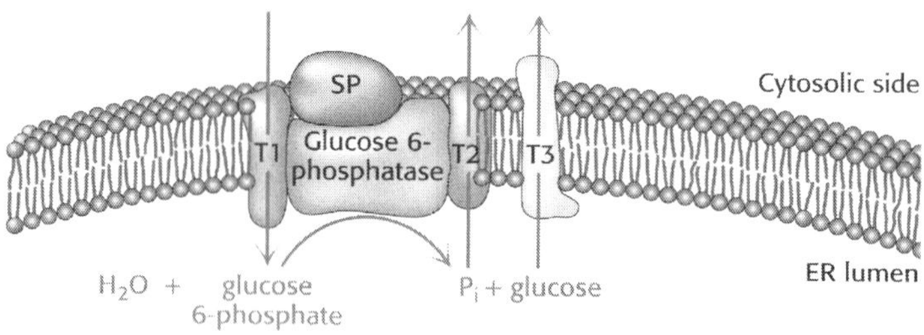

The lumen of the ER: It takes 5 proteins to convert glucose 6-phosphate to glucose.

Gluconeogenesis Stoichiometry

2pyruvate + 4ATP + 2GTP + 2NADH + $6H_2O \rightarrow$ glucose + 4ADP + 2GDP $+6P_i + 2NAD^+ + 2H^+$; G°' = -9kcal mol^{-1}.

Just the reverse of glycolysis, G°' = 20kcal mol^{-1}. Note: it takes 6 nucleotide triphosphate molecules to synthesize glucose. Only 2 nucleotide triphosphate molecules are generated from glycolysis. So it takes four extra high phosphoryl-transfer potential molecules to drive the unfavorable gluconeogenesis pathway.

Glycolysis and gluconeogenesis are reciprocally regulated

A high [AMP] indicates that the energy charge is low and signals the need for ATP. High [ATP] and [citrate] indicate the energy charge is high and intermediates are abundant.

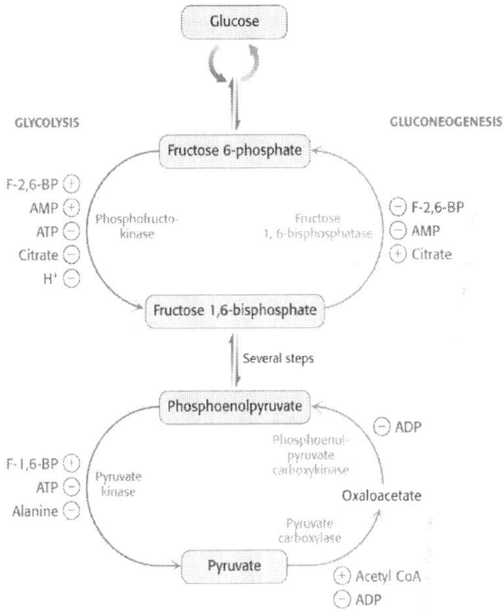

The Cori Cycle

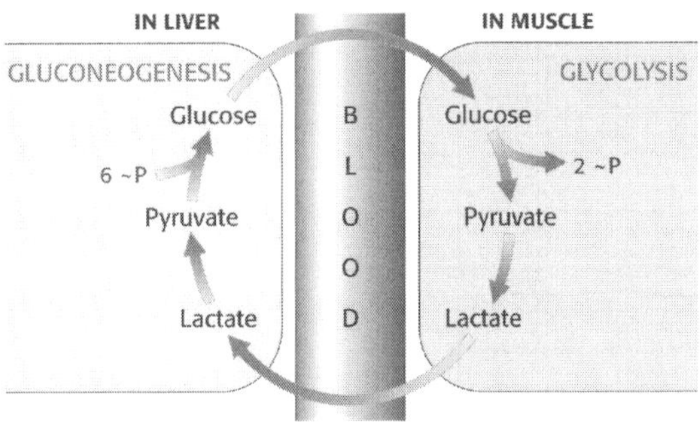

Lactate from active muscle is converted to glucose in the liver. **Carl and Gerty Cori** won the Nobel Prize in Physiology and medicine 1947: "for their discovery of the course of the catalytic conversion of glycogen"

Lactate and Alanine are Glucogenic

In muscles, alanine is produced from pyruvate by transamination. In the liver, alanine is converted back to pyruvate. In active muscles, lactate builds up, passes through the blood and is converted to pyruvate in the liver. Thus, part of the metabolic burden of active muscle is shifted to the liver.

pyruvate + glutamate ⇆ alanine + α-ketoglutarate

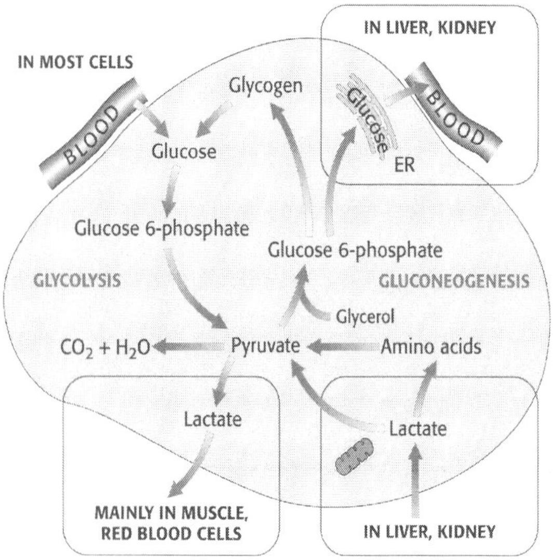

Gluconeogenesis

The source of pyruvate and oxaloacetate for gluconeogenesis during fasting or carbohydrate starvation is mainly amino acid catabolism. Some amino acids are catabolized to pyruvate, oxaloacetate, or precursors of these. Muscle proteins may break down to supply amino acids. These are transported to the liver where they are deaminated and converted to gluconeogenesis inputs. Glycerol, derived from hydrolysis of triacylglycerols in fat cells, is also a significant input to gluconeogenesis.

Gluconeogenesis is the process whereby precursors such as lactate, pyruvate, glycerol, and amino acids are converted to glucose. Fasting requires all the glucose to be synthesized from these non-carbohydrate precursors. Most precursors must enter the Krebs cycle at some point to be converted to oxaloacetate. Oxaloacetate is the starting material for gluconeogenesis.

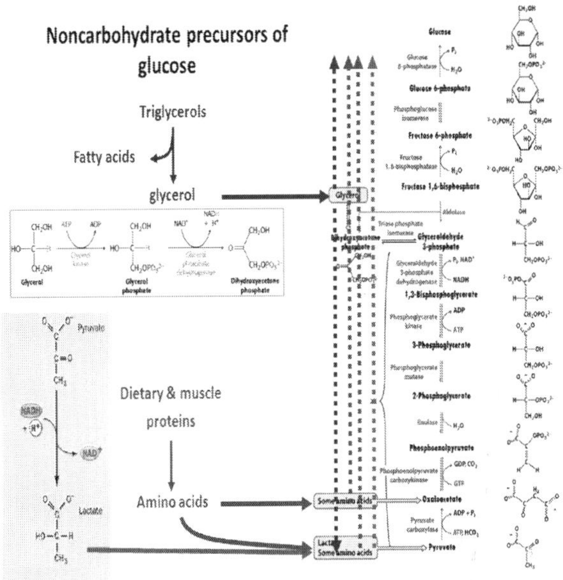

Main sites of gluconeogenesis Major site (Liver), Minor site: (Kidney) Very little: (Brain, Muscle (skeletal and heart). In the liver and kidney, it helps to maintain the glucose level in the blood so that brain and muscles can extract sufficient glucose from it to meet their metabolic demands.

Gluconeogenesis versus Glycolysis

- steps are shared between glycolysis and gluconeogenesis. 3 essentially irreversible steps shift the equilibrium far on the side of glycolysis.
- Most of the decrease in free energy (consuming energy) in glycolysis takes place during these 3 steps.

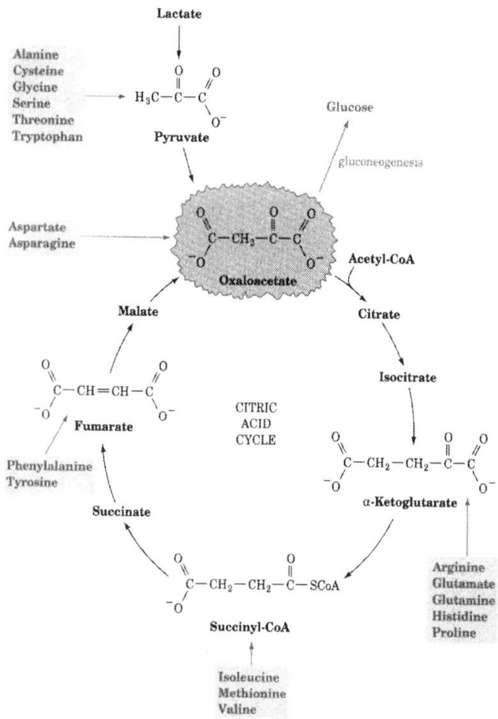

Pyruvate Is Converted To Oxaloacetate Before Being Changed To Phosphoenolpyruvate

1. Pyruvate carboxylase catalyzes the ATP-driven formation of oxaloacetate from pyruvate and CO_2
2. PEP carboxykinase (PEPCK) concerts oxaloacetate to PEP that uses GTP as a phosphorylating agent.

Pyruvate Carboxylase Requires Biotin As A Cofactor

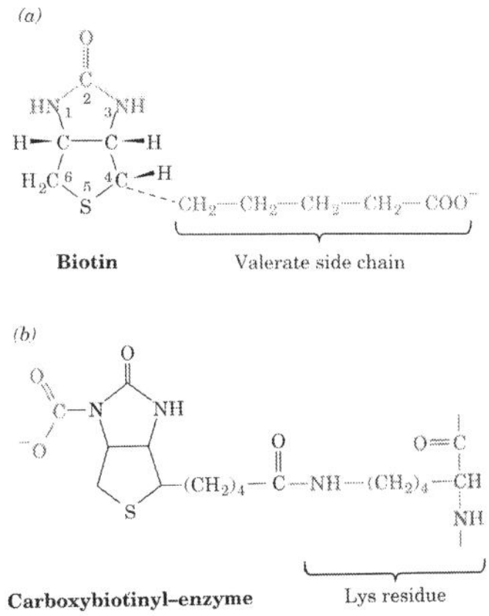

Gluconeogenesis Is Not Just The Reverse Of Glycolysis

Several steps are different so that control of one pathway does not inactivate the other. However, many steps are the same. Three steps are different from glycolysis.

1. Pyruvate to PEP
2. Fructose 1,6- bisphosphate to Fructose-6-phosphate
3. Glucose-6-Phosphate to Glucose.

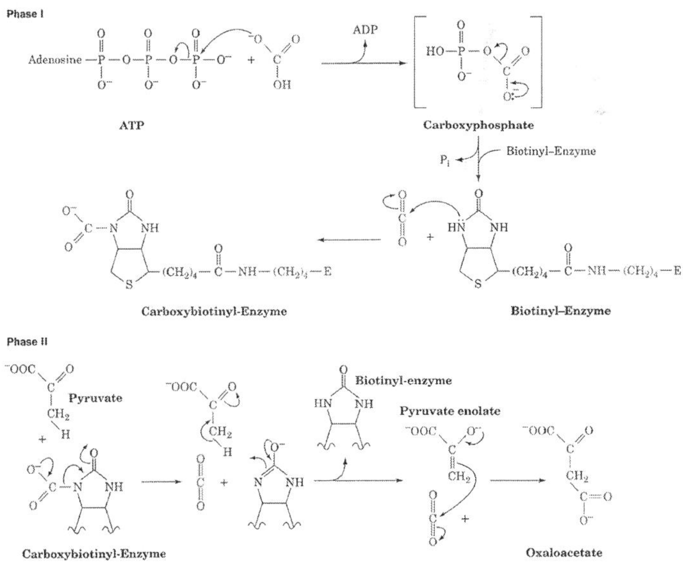

BIOTIN IS AAN ESSENTIAL NUTRIENT

There are hardly any deficiencies for biotin because it is abundant and bacteria in the large intestine also make it. However, deficiencies have been seen and are nearly always linked to the consumption of raw eggs. Raw eggs contain Avidin a protein that binds biotin with a $K_d = 10^{-15}$ (that is one tight binding reaction!) It is thought that Avidin protects eggs from bacterial invasion by binding biotin and killing bacteria.

PEP Carboxykinase

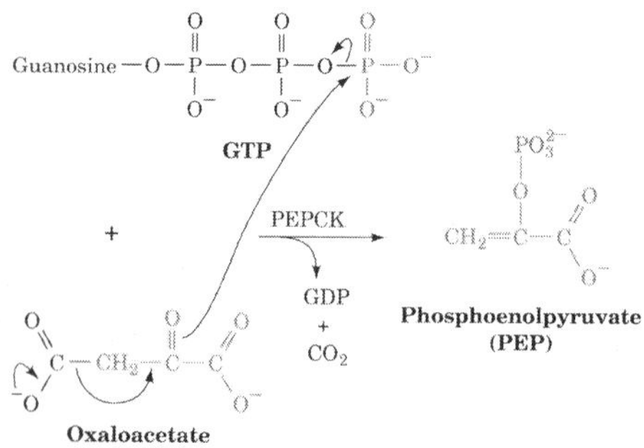

Acetyl-Coa Regulates Pyruvate Carboxylase

Oxaloacetate concentrations increase then increase the activity of the Krebs cycle and Acetyl-CoA is an allosteric activator of the carboxylase. However, when ATP and NADH concentrations are high and the Krebs cycle is inhibited, oxaloacetate goes to glucose.

Transport between the Mitochondria and the Cytosol

Generation of oxaloacetate occurs in the mitochondria only, but, gluconeogenesis occurs in the cytosol. PEPCK is distributed between both compartments in humans, while in mice, it is only found in the cytosol. In rabbits, it is found in the mitochondria. PEP must be transported across the membranes or oxaloacetate has to be transported. PEP transport systems are seen in the mitochondria but oxaloacetate cannot be transported directly in or out of the mitochondria.

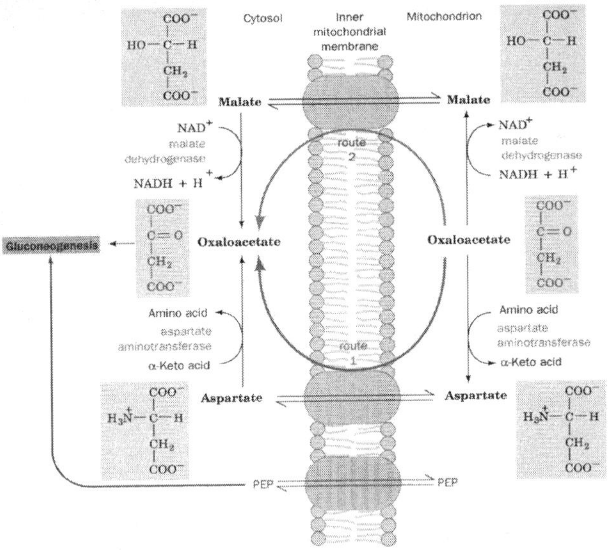

Hydrolytic Reactions Bypass PFK and Hexokinase

The hydrolysis of fructose-1,6-phosphate and glucose-6- phosphate are separate enzymes from glycolysis. Glucose-6-phosphatase is only found in the liver and kidney. The liver is the primary organ for gluconeogenesis.

Glucose + $2NAD^+$ + 2ADP + 2Pi

2Pyruvate +2NADH + $4H^+$ + 2ATP + $2H_2O$

2Pyruvate +2NADH + $4H^+$ + 4ATP + 2GTP + $6H_2O$

glucose + $2NAD^+$ + 4ATP + 2GDP + 4Pi

2ATP + 2GTP + $4H_2O$ 2ADP + 2GTP + 4Pi

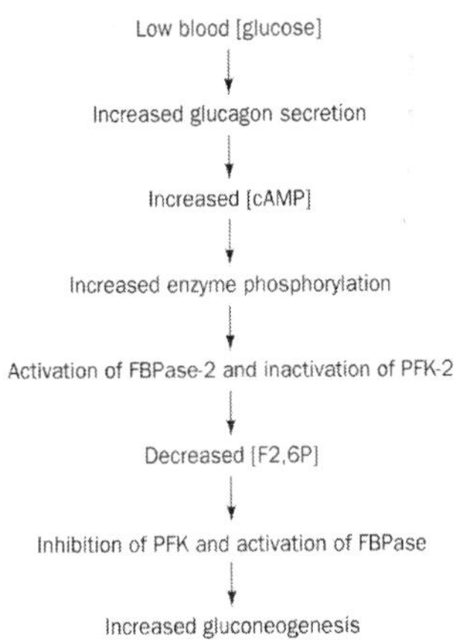

The Glyoxylate Pathway

Only plants have the ability to convert Acetyl-CoA to Oxaloacetate directly without producing reducing equivalents of NADH. This is done in the glyoxysome, separate from the mitochondria and allows replenishment of oxaloacetate. Isocitrate lyase - cleaves isocitrate into succinate and glyoxylate. The succinate goes to the mitochondria. Malate synthase makes malate from glyoxylate and Acetyl-CoA. The oxaloacetate can go directly to carbohydrate synthesis.

Glycogen Storage

Glycogen is a D-glucose polymer, a(1→4) linkages, a(1→6) linked branches every 8-14 residues

The Citric Acid Cycle

It is called the Krebs cycle or the tricarboxylic and is the "hub" of the metabolic system. It accounts for the majority of carbohydrate, fatty acid and amino acid oxidation. It also accounts for a majority of the generation of these compounds and others as well. Amphibolic - acts both catabolically and anabolically

3NAD+ + FAD + GDP + Pi + acetyl-CoA 3NADH + FADH + GTP + CoA + $2CO_2$

By 1930 it was established that the addition of lactate, acetate succinate, malate, a-ketoglutaric acid (dicarboxylic acids) and citrate and isocitrate (tricarboxylic acids) when added to muscle mince that they stimulated oxygen consumption and release of CO_2. Albert Szent-Gyorgyi (1935) showed that

Succinate → Fumarate → Malate → Oxaloacetate

Carl Martius and Franz Knoop Showed

Citrate → cis-aconitate→ isocitrate → a ketoglutarate

succinate → fumarate → malate→ oxaloacetate

Martius and Knoop showed that pyruvate and oxaloacetate could form citrate non-enzymatically by the addition of peroxide under basic conditions. Krebs showed that succinate is formed from fumarate, malate or oxaloacetate. This is interesting since it was shown that the other way worked as well. Pyruvate can form citrate enzymatically. The interconversion rates of the intermediates were fast enough to support respiration rates.

Pyruvate + oxaloacetate = citrate + CO_2

The citric acid cycle enzymes are found in the matrix of the mitochondria. Substrates have to flow across the outer and inner parts of the mitochondria. CoA acts as a carrier of acetyl groups. Acetyl-CoA is a "high energy" compound: The DG°' for the hydrolysis of its thioester is -31.5 kJ• mol^{-1} making it greater than the hydrolysis of ATP. Pyruvate dehydrogenase converts pyruvate to Acetyl-CoA and CO_2.

Pyruvate Dehydrogenase

Multienzyme complexes are groups of non-covalently associated enzymes that catalyze two or more sequential steps in a metabolic pathway.

1. Enzymatic reactions rates are limited by diffusion, with shorter distance between subunits of an enzyme almost directing the substrate from one subunit (catalytic site) to another.
2. Channeling metabolic intermediates between successive enzymes minimizes side reactions
3. The reactions of a multienzyme complex can be coordinately controlled

Pyruvate Dehydrogenase

- The enzyme decarboxylates pyruvate using a TPP cofactor forming hydroxyethyl-TPP.
- The hydroxyethyl group is transferred to the oxidized lipoamide on E2 to form Acetyl dihydrolipoamide-E2.
- E2 catalyzes the transfer of the acetyl groups to CoA yielding Acetyl-CoA and reduced dihydrolipoamide-E2.
- Dihydrolipoyl dh E3 reoxidizes dihydrolipoamide-E2 and itself becomes reduced as FADH2 is formed.
- Reduced E3 is reoxidized by NAD^+ to form FAD and NADH The enzymes SH groups are reoxidized by the FAD and the electrons are transferred to NADH.

Electron Transport Chain

In eukaryotes => Electron transport and oxidative phosphorylation => inner **mitochondrial** membrane. These processes => re-oxidize NADH and FADH2 <= from the citric acid cycle (mitochondrial matrix), glycolysis (cytoplasm) and fatty acid oxidation (mitochondrial matrix) and => trap the energy released as ATP.

Oxidative phosphorylation => major source of ATP in the cell.

In prokaryotes => electron transport and oxidative phosphorylation components => in the plasma membrane.

Redox Potential

Oxidation => loss of electrons.

Reduction => gain of electrons.

In chemical reaction: if one molecule is oxidized => another must be reduced i.e. oxidation-reduction reaction => transfer of electrons. When NADH => oxidized to NAD+ => it loses electrons. When molecular oxygen => reduced to water => it gains electrons. Oxidation-reduction potential, E, (redox potential). a measure of the affinity of a substance for electrons and is measured relative to hydrogen.

A **Positive** redox potential substance => higher affinity => electrons than hydrogen so would accept electrons from hydrogen, e.g., Oxygen, a strong oxidizing agent.

A **Negative** redox potential substance has a lower affinity for electrons than does hydrogen would donate electrons to H+, forming hydrogen e.g., NADH, a strong reducing agent.

For Biological Systems

Total voltage changes of the reaction (change in electric potential, ΔE) => is the sum of voltage changes of individual oxidation-reduction steps. **Standard free energy change** of a reaction at pH 7 => **ΔG0'**=> calculated from the change in redox potential ΔE0' of substrates and products:

ΔG0' = -n F ΔE0'

Where, n = number of electrons transferred,

ΔE0' in volts (V),

ΔG0' in kilocalories per mole (kcal mol-1) and

F-constant called Faraday (23.06 kcal V-1 mol-1).

A reaction with a positive ΔE0' has a negative ΔG0' (i.e., is **exergonic**). Thus for the reaction

Electron Transport from NADH

Comparing the energetic of the oxidation of NADH:

$$NADH + H^+ + \frac{1}{2} O_2 \rightleftharpoons NAD^+ + H_2O \qquad \Delta G^{0'} = -52.6 \text{ kcal mol}^-$$

and the synthesis of ATP:

$$ADP + P_i + H^+ \rightleftharpoons ATP + H_2O \qquad \Delta G^{0'} = +7.3 \text{ kcal mol}^-$$

Thus, the oxidation of NADH releases sufficient energy to drive the synthesis of several molecules of ATP.

- NADH oxidation and ATP synthesis → not occur in a single step.
- Electrons → not transferred from NADH → oxygen directly.
- Electrons are transferred from NADH → oxygen → along a chain of electron carriers → called **electron transport chain (respiratory chain)**.

Electron Transport Chain

It consists of 3 large protein complexes embedded in the inner mitochondrial membrane. NADH dehydrogenase complex (Complex I). Succinate Q reductase. The cytochrome bc1 complex (Complex II) cytochrome oxidase (Complex III). Electrons flow from NADH to oxygen through these three complexes. Each complex contains several electron carriers that work sequentially carrying electrons down the chain. There are 2 free electron carriers are also needed to link these large complexes.

Ubiquinone {coenzyme Q (CoQ)}

Cytochrome c

ATP Synthesis (Oxidative Phosphorylation)

NADH and FADH2 are oxidized by electron transport through the respiratory chain: synthesis of ATP. The energy liberated by electron transport is used to create a proton gradient across the mitochondrial inner membrane that is used to drive ATP synthesis (chemiosmotic hypothesis) in presence of ATP synthase. Thus the proton gradient couple's electron transport and ATP synthesis.

(not a chemical intermediate as in substrate level phosphorylation.)

Summary

Electron transport down the respiratory chain from NADH oxidation causes H+ ions to be pumped out into the intermembrane space by three H+ pumps NADH dehydrogenase, cytochrome bc1 complex and cytochrome oxidase. Free energy change in transporting an electrically charged ion across a membrane leads to the formation of an electrochemical proton gradient. The pumping out of H+ ions generates a higher concentration of H+ ions in intermembrane space and an electrical potential on the side of the inner mitochondrial membrane facing the intermembrane space positive. Protons flow back into mitochondrial matrix according to electrochemical gradient through ATP synthase drives ATP synthesis.

The ATP synthase is driven by proton-motive force which is the sum of pH gradient (the chemical gradient of H+ ions) and membrane potential (electrical charge potential across the inner mitochondrial membrane). **FADH2** is reoxidized via ubiquinone its oxidation causes H+ ions to be pumped out only by the **cytochrome bc1 complex and cytochrome oxidase** so the amount of ATP made from FADH2 is less than from NADH. Measurements show that 2.5 ATP molecules are synthesized per NADH oxidized whereas 1.5 ATPs are synthesized per FADH2 oxidized.

Oxidative Phosphorylation

Carbohydrates, lipids, and amino acids are the main reduced fuels for the cell. Electrons from reduced fuels are transferred to reduced cofactors NADH or $FADH_2$. In oxidative phosphorylation, energy from NADH and $FADH_2$ is used to make ATP.

Chemiosmotic Theory

ADP + $P_i \rightarrow$ ATP is highly thermodynamically unfavorable.

Phosphorylation of ADP is not a result of a direct reaction between ADP and some high-energy phosphate carrier. The energy needed to phosphorylate ADP is provided by the flow of protons down the electrochemical gradient. The energy released by electron transport is used to transport protons against the electrochemical gradient.

Chemiosmotic Energy Coupling Requires Membranes

The proton gradient needed for ATP synthesis can be stably established across a membrane that is impermeable to ions; plasma membrane in bacteria, inner membrane in mitochondria and thylakoid membrane in chloroplasts. The membrane must contain proteins that couple the "downhill" flow of electrons in the electron-transfer chain with the "uphill" flow of protons across the membrane. The membrane must contain a protein that couples the "downhill" flow of protons to the phosphorylation of ADP.

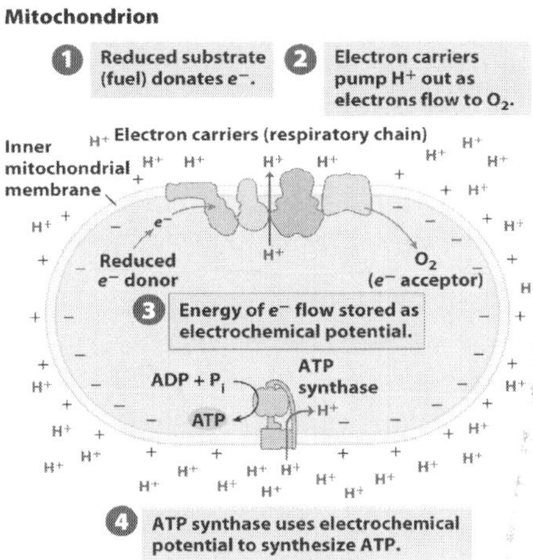

Figure 19-1
Lehninger Principles of Biochemistry, Seventh Edition
© 2017 W. H. Freeman and Company

Structure of a Mitochondrion

Double membrane leads to four distinct compartments:

- ***Outer membrane:*** Relatively porous membrane; allows passage of metabolites.
- ***Intermembrane space (IMS):*** Similar environment to cytosol and higher proton concentration (lower pH).

Inner membrane: Relatively impermeable, with proton gradient across it, Location of electron transport chain complexes and convolutions called cristae serve to increase the surface area.

Matrix: Location of the citric acid cycle and parts of lipid and amino acid metabolism and lower proton concentration (higher pH)

Electron-Transport Chain Complexes Contain a Series of Electron Carriers

Each complex contains multiple redox centers consisting of: Flavin mononucleotide (FMN) or flavin adenine dinucleotide (FAD), Cytochromes *a*, *b*, or *c* and Iron-sulfur cluster.

Complex I

Complex I (shown here is the crystal structure from the bacterium *Thermus thermophilus*) catalyzes the transfer of a hydride ion from NADH to FMN, and from the FMN, two electrons pass through a series of FeScenters to the Fe-S Center N-2 in the matrix arm of the complex. Electron transfer from N-2 to ubiquinone on the membrane arm forms QH2, which diffuses into the lipid bilayer. This electron transfer also drives expulsion from the matrix of four protons per pair of electrons. Proton flux produces an electrochemical potential across the inner mitochondrial membrane (N side negative, P side positive). The detailed mechanism that couples electron and proton transfer in Complex I is not yet known, but three of the membrane subunits are structurally related to a known Naƨ-Hƨanti-porter, and the path of proton movement may be similar in both cases. The fourth putative proton pathway is through an integral subunit closer to the Q-binding site. A long helix (not visible in this view) lying along the surface of the membrane arm may coordinate the activities of all four proton pumps when Q is reduced.

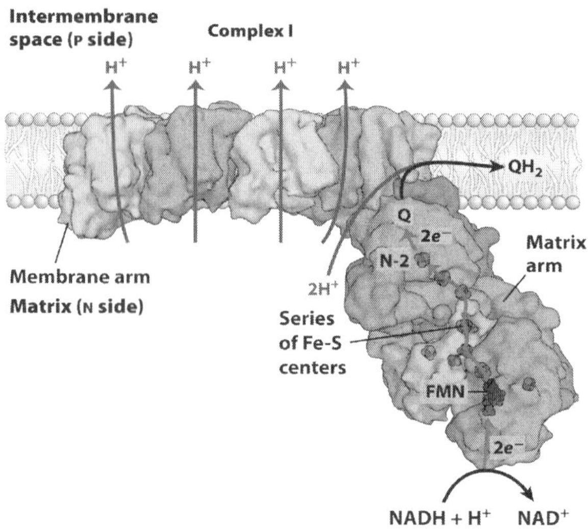

Figure 19-8
Lehninger Principles of Biochemistry, Seventh Edition
© 2017 W. H. Freeman and Company

Complex II

This complex (porcine) has two transmembrane subunits, C and D; subunits A and B extend into the matrix. Just behind the FAD in subunit A is the binding site for succinate. Subunit B has three Fe-S centers, ubiquinone is bound to subunit B, and heme b is sandwiched between subunits C and D.

Two phosphatidylethanolamine molecules are so tightly bound to sub-unit D that they show up in the crystal structure. Electrons move (blue arrows) from succinate to FAD, then through the three Fe-S centers to ubiquinone. The heme b is not on the main path of electron transfer but protects against the formation of reactive oxygen species (ROS) by electrons that go astray.

Complex III

The complex (bovine) is a dimer of identical monomers, each with 11 different subunits. The functional core of each monomer consists of three subunits: cytochrome b (green), with its two hemes (bH and bL); the Rieske iron-sulfur protein (purple), with its 2Fe-2S centers; and cytochrome c1 (blue), with its heme. This cartoon illustration shows how cytochrome c1 and the Rieskeiron-sulfur protein project from the P surface and can interact with cytochrome c (not part of the functional complex) in the intermembrane space. The complex has two distinct binding sites for ubiquinone, QN and QP, which correspond to the sites of inhibition by two drugs that block oxidative phosphorylation. Antimycin A, which blocks electron flow from cytochrome b to cytochrome c1, specifically from heme bH to Q, binds at QN, close to heme bH on the N (matrix) side of the membrane. Myxothiazol, which prevents selectron flow from QH2 to the Rieske iron-sulfur protein, binds at QP, near the2Fe-2S center and heme bL on the P side. The dimeric structure is essential to the function of Complex III. The interface between monomers forms two caverns, each containing a QP site from one monomer and a QN site from the other. The ubiquinone intermediates move within these sheltered caverns. Complex III crystallizes in two distinct conformations (not shown). In one, the Rieske Fe-S center is close to its electron acceptor, the heme of cytochrome c1, but relatively distant from cytochrome b and the QH2-binding site at which the Rieske Fe-S center receives electrons. In the other, the Fe-S center has moved away from cytochrome c1 and toward cytochrome b. The Rieske protein is thought to oscillate between these two conformations as it is first reduced, then oxidized.

Complex IV

This complex (bovine) has 13 subunits in each identical monomer of its dimeric structure. Subunit I (yellow) has two heme groups, a and a3, near a single copper ion, CuB (not visible here). Heme a3 and CuB form a binuclear Fe-Cu center. Subunit II (purple) contains two Cu ions complexed with the OSH groups of two Cys residues in a binuclear center, CuA, that resembles the 2Fe-2S centers of iron-sulfur proteins. This binuclear center and the cytochrome c–binding site are located in a domain of subunit II that protrudes from the P side of the inner membrane (into the intermembrane space). Subunit III (blue) is essential for rapid proton movement through subunit II. The roles of the other 10subunits in mammalian Complex IV (green) are not fully understood, although some function in assembly or stabilization of the complex. The binuclear center of CuA. The Cu ions (blue spheres) share electrons equally. When the center is reduced, the ions have the formal charges Cu11Cu11; when oxidized, Cu1.51Cu1.51. Six amino acid residues are ligands around the Cu ions: Glu, Met, two His, and two Cys.

Summary of the Electron Flow in the Respiratory Chain

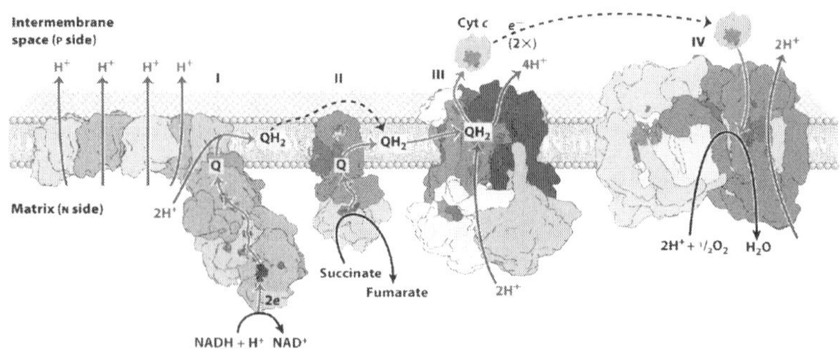

Figure 19-16
Lehninger Principles of Biochemistry, Seventh Edition
© 2017 W. H. Freeman and Company

Proton-Motive Force

The proteins in the electron-transport chain created the electrochemical proton gradient by one of three means: actively transporting protons across the membrane Complex I and Complex IV, chemically removing protons from the matrix, reduction of CoQ and reduction of oxygen, releasing protons into the intermembrane space and oxidation of QH_2.

Chemiosmotic Model for ATP Synthesis

Electron transport sets up a proton-motive force. The energy of proton-motive force drives the synthesis of ATP.

Mitochondrial ATP Synthase Complex

It contains two functional units: F_1: soluble complex in the matrix and individually catalyzes the hydrolysis of ATP. F_0: integral membrane complex, transports protons from IMS to the matrix, dissipating the proton gradient, the energy transferred to F_1 to catalyze the phosphorylation of ADP.

Regulation of Oxidative Phosphorylation

Primarily regulated by substrate availability NADH and ADP/P_i. **Inhibitor of F_1 (IF_1):** prevents hydrolysis of ATP during low oxygen and only active at lower pH, encountered when electron transport it stalled (i.e., low oxygen). Inhibition of OxPhos leads to accumulation of NADH. It causes feedback inhibition to cascade up to PFK-1 in glycolysis.

Interrelations Helps of Metabolic Pathway

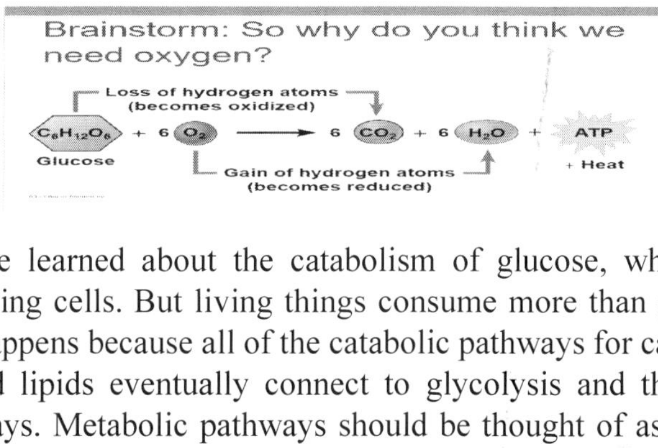

You have learned about the catabolism of glucose, which provides energy to living cells. But living things consume more than glucose from food. This happens because all of the catabolic pathways for carbohydrates, proteins, and lipids eventually connect to glycolysis and the citric acid cycle pathways. Metabolic pathways should be thought of as porous; that is, substances enter from other pathways, and intermediates leave for other pathways. These pathways are not closed systems. Many of the substrates, intermediates, and products in a particular pathway are reactants in other pathways. Like sugars and amino acids, the catabolic pathways of lipids are also connected to the glucose catabolism pathways. Glycogen, a polymer of glucose, is an energy-storage molecule in animals. When there is adequate ATP present, excess glucose is shunted into glycogen for storage. Glycogen is made and stored in both the liver and muscles. The glycogen is hydrolyzed into the glucose monomer, glucose-1-phosphate (G-1-P), if blood sugar levels drop. The presence of glycogen as a source of glucose allows ATP to be produced for a longer period of time during exercise. Glycogen is broken down into G-1-P and converted into glucose-6-phosphate (G-6-P) in both muscle and liver cells; this product enters the glycolytic pathway.

Excess amino acids are converted into molecules that can enter the pathways of glucose catabolism. Metabolic pathways should be thought of as porous; that is, substances enter from other pathways and intermediates leave for other pathways. These pathways are not closed systems. Many of the substrates, intermediates, and products in a particular pathway are reactants in other pathways. Proteins are a good example of this phenomenon. They can be broken down into their constituent amino acids and used at various steps of the pathway of glucose catabolism. Proteins are hydrolyzed by a variety of enzymes in cells. Most of the time, the amino acids are recycled into the synthesis of new proteins or are used as precursors in the synthesis of other important biological molecules, such as hormones, nucleotides, or neurotransmitters. However, if there are excess amino acids, or if the body is in a state of starvation, some amino acids will be shunted into the pathways of glucose catabolism. Each amino acid must have its amino group removed (deamination) prior to the carbon chain's entry into these pathways. When the amino group is removed from an amino acid, it is converted into ammonia through the urea cycle. The remaining atoms of the amino acid result in a keto acid: a carbon chain with one ketone and one carboxylic acid group. In mammals, the liver synthesizes urea from two ammonia molecules and a carbon dioxide molecule. Thus, urea is the principal waste product in mammals produced from the nitrogen originating in amino acids; it leaves the body in urine. The keto acid can then enter the citric acid cycle. When deaminated, amino acids can enter the pathways of glucose metabolism as pyruvate, acetyl CoA, or several components of the citric acid cycle. For example, deaminated asparagine and aspartate are converted into oxaloacetate and enter glucose catabolism in the citric acid cycle.

Connecting Lipids to Glucose Metabolism

Lipids can be both made and broken down through parts of the glucose catabolism pathways. Like sugars and amino acids, the catabolic pathways of lipids are also connected to the glucose catabolism pathways. The lipids that are connected to the glucose pathways are cholesterol and triglycerides. Cholesterol contributes to cell membrane flexibility and is a precursor to steroid hormones. The synthesis of cholesterol starts with acetyl groups, which are transferred from acetyl CoA, and proceeds in only one direction; the process cannot be reversed. Thus, the synthesis of cholesterol requires an intermediate of glucose metabolism.

Triglycerides

Triglycerides, a form of long-term energy storage in animals, are made of glycerol and three fatty acids. Animals can make most of the fatty acids they need. Triglycerides can be both made and broken down through parts of the glucose catabolism pathways. Glycerol can be phosphorylated to glycerol-3-phosphate, which continues through glycolysis. Fatty acids are catabolized in a process called beta-oxidation that takes place in the matrix of the mitochondria and converts their fatty acid chains into two-carbon units of acetyl groups while producing NADH and $FADH_2$. The acetyl groups are picked up by CoA to form acetyl CoA that proceeds into the citric acid cycle as it combines with oxaloacetate. The NADH and $FADH_2$ are then used by the electron transport chain.

Biochemistry of Seed Germination and Development

It provides protection against entry of parasites, it protects against mechanical injury, in some seeds, it buffers against unfavorably high or low temperatures. It stops germination until the right time. Inside of the seed coat, an embryo is protected. The embryo is an immature plant with all of the parts of an adult plant. A close look shows leaves and roots, albeit very small leaves and roots. The leaves of the embryo are called plumules. The leaves are sheathed by a cotyledon. The embryonic roots are called radicles. The embryonic stem is called the hypocotyl. The seed is filled with nourishment for the new plant; this source of nutrition is called the endosperm.

Germination

The first step of germination is the absorption of water a lot of water. The absorption of water activates enzymes in the seed that stimulate growth. These enzymes break down starches in the endosperm into sugars that can be used for energy. The deciding factor in whether or not a seed germinates is whether or not energy is available for growth and cell division.

Stages Leading to Cell Division

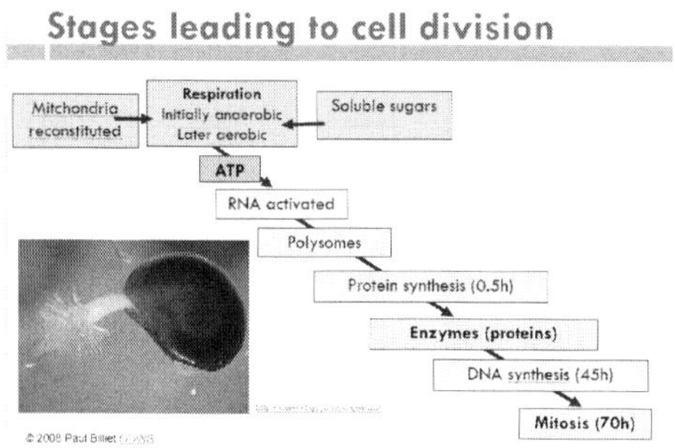

Seed Dormancy

If moisture is the key to starting seed germination, it should be clear what is necessary for seed dormancy, a waterproof seed coat, no oxygen and chemical inhibitors. For germination to occur, these must be negated, coat broken down; oxygen available and water absorbed and growth promoters activated, inhibitors are inhibited.

Germination

Stage	Events
Pregermination	(a) Rehydration – imbibition of water. (b) RNA & protein synthesis stimulated. (c) Increased metabolism – increased respiration. (d) Hydrolysis (digestion) of food reserves by enzymes. (e) Changes in cell ultrastructure. (f) Induction of cell division & cell growth.
Germination	(a) Rupture of the seed coat. (b) Emergence of seedling, usually radicle first.
Post germination	(a) Controlled growth of root and shoot axis. (b) Controlled transport of materials from food stores to the growing axis. (c) Senescence (aging) of food storage tissues.

Germination Factors

Water is clearly the most important factor in germination an adequate continuous supply of water is necessary for complete emergence. Water functions as a triggering enzyme for starch conversion into sugar, turgor pressure for moving the radicle root down and the cotyledons up and for transporting nutrients and enzymes within the seed. Light is another key germination factor; light can either stimulate or inhibit seed germination. Some crops have a requirement for light to assist seed germination (eg. begonias, impatiens, lettuce). Others germinate best in the dark. This is determined by how the seed would naturally be sown. Small seeds must sprout on the surface of soil because they lack a suitable endosperm to supply the needed nutrients; these are typically aided by light exposure. Large seeds contain enough nutrition to grow underground when photosynthesis is not possible. These seeds are more likely to germinate in dark conditions. Oxygen is the third factor. Cellular respiration is necessary for plants to grow; oxygen is necessary to complete respiration. Oxygen removes metabolic waste from the cell. Waste is not removed without oxygen and the cellular metabolism is slowed. If oxygen supply is limited during germination, emergence may not occur due to inhibited growth. A favorable temperature is necessary to allow for plant growth. Temperature not only affects the germination percentage but also the rate of germination. For every species of seed, there is an optimal soil temperature for germination. At that temperature, the maximum number of seeds will germinate and in less time than at any other temperature. Many seeds germinate best around room temp. The germination is a complex process. A pre-formed plant (embryo) inside of the seed coat must turn the endosperm (starch) into sugar. This sugar powers cell division (mitosis); the addition of cells will cause the embryonic roots, leaves, and stems to grow, expand, and develop. During germination, the radicle (embryonic root) emerges due to mitosis fueled by the breakdown of starch into sugar. Under warm conditions, this process will take 4-5 days. Under cool conditions, this process takes longer. Initially, the radicle grows in what direction the kernel tip is pointing. Later, smaller roots will emerge from the radicle at varying angles. These roots will absorb the nutrients necessary for growth and development when the endosperm is completely consumed.

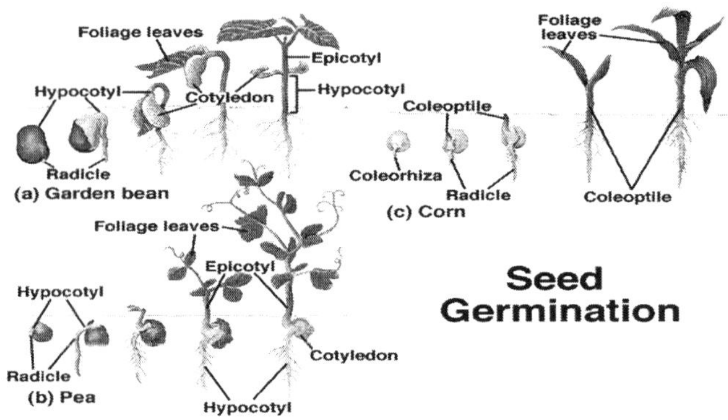

Adapted from Randy Moore, Dennis Clark, and Darrell Vodopich, Botany Visual Resource Library, MacGraw-hill companies, Inc. (1998).

The coleoptile, a protective sheath covering the shoot, pushes through the soil until it reaches the light. Upon reaching light, the plumule emerges. If the kernel is planted too deep, emergence will not occur. The time between planting and emergence is determined mostly by temperature; warmer temps are reduced emergence time. This is only true to an extent; excessively high temps can also increase emergence time. Soil compaction can increase the time to emergence. Lack of moisture can also increase emergence time.

Sprouting Time

Many temperate-zoned species use chemical inhibitors to induce dormancy in seeds. In fall the seeds are exposed to abscisic acid (ABA) by the plant to prevent the seeds from sprouting before winter. Over winter, enzymes in the seed degrade the abscisic acid and by spring it is gone.

Vernalization

Seeds of some species can be caused to sprout earlier by exposure to cool or cold temperatures. Seeds of temperate species that were exposed to the cold will sprout earlier in a greenhouse than those planted directly. This process is called vernalization. Vernalization is the intentional exposing of seeds to cool conditions to increase the breakdown of chemical inhibitors and stimulate the production of growth or flowering enzymes. Many annuals are facultative and vernalization is not necessary for development but does speed it up.

Biochemistry of Fruit Ripening

Ripening is the process by which fruits attain their desirable flavor, quality, color, palatable nature and other textural properties. Ripening is associated with change in composition i.e. conversion of starch to sugar. On the basis of ripening behavior, fruits are classified as climacteric and non-climacteric fruits. Climacteric: Climacteric fruits are defined as fruits that enter 'climacteric phase' after harvest i.e. they continue to ripen. During the ripening process, the fruits emit ethylene along with an increased rate of respiration. Ripe fruits are soft and delicate and generally cannot withstand rigors of transport and repeated handling. These fruits are harvested hard and green, but fully mature and are ripened near consumption areas. A small dose of ethylene is used to induce the ripening process under controlled conditions of temperature and humidity. These fruits in a fully ripe state are too delicate to withstand transportation over long distances and should preferably be ripened near the consumption area. Non-climacteric: Non-climacteric fruits once harvested do not ripen further. Non-climacteric fruits produce a very small amount of ethylene and do not respond to ethylene treatment. There is no characteristic increased rate of respiration or production of carbon dioxide.

Technologies for ripening of fruits ack of easier and rapid methods for uniform ripening poses a major problem in the fruit industry. "Almost all methods of ripening, either conventional or the modern chemical methods, come with their own merits and demerits. There are several simple technologies and methods available today for farmers for proper ripening. Normally, the number of days taken for edible ripening varies for different fruits and prevailing climatic conditions. For instance, it takes about 5 to 6 days for mangoes and 6 to 7 days for sapotas to ripen. Under natural conditions, ethylene, a ripening hormone produced by the plant plays a major physiological role in the ripening process. A simple technology practiced in households to trigger ripening is to keep unripened and ripened fruits together inside an air tight container. Since the already ripened fruits release ethylene, ripening will be faster. Another method is to place the fruits intended for ripening inside an air tight room and induce ripening through smoking inside smoke chambers. Smoke emanates acetylene gas. Several fruit traders follow this technique to achieve uniform ripening especially in edible fruits like banana and mango. But the major drawback of this method is that the fruits do not attain uniform color and flavor. In addition, the persistence of smoke odor on the product impairs its quality. Spreading unripe fruits as layers over paddy husk or wheat straw for a week to ripen is an alternative. Another practice is that some farmers dip unripe mature fruits in 0.1 percent ethrel solution (1 ml of ethrel solution in 1 liter of water) and wipe it dry. The fruits are then spread over a newspaper without touching each other and a thin cotton cloth is covered over this. In this method, the fruits will ripen within two days. In one of the simple and harmless techniques, 10 ml of ethrel and 2 gm of sodium hydroxide pellets are mixed in five liters of water taken in a wide-mouthed vessel. This vessel is placed inside the ripening chamber near the fruits and the room is sealed air tight. About a third of the room is filled with fruits leaving the remaining area for air circulation. Ripening of fruits takes place in about 12 to 24 hours. In order to reduce the cost of chemicals, some ethylene releasing fruits such as papaya and banana can also have kept in the same room". Ethylene gas filled in pressurized cans promote fruit ripening in 24-48 hours.

Chapter 4

ELEMENTS OF IMMUNOLOGY

29. INTRODUCTION TO IMMUNOLOGY

Introduction

It is the science that is concerned with the immune response to foreign challenges or simply the study of the body's defense against infection. It addresses the questions such as how does the body defend itself against infection, when an infection does occur, how does the body eliminate the pathogens and how does long-lasting immunity to many infectious diseases develop? The ability of an organism to resist infections by pathogens or a state of protection against foreign organisms or substances is called immunity.

The array of cells, tissues and organs which carry out this activity constitute the immune system. The immune system is a complex process and it is derived into two categories such as innate (native) and adaptive (acquired) immunity.

Innate Immunity: It is a general, non-specific immune response that presents in all individuals at all times.

Adaptive immunity: It is highly specific to the particular pathogen that induces it.

30. IMMUNE SYSTEMS-OVERVIEW, IMMUNITY WITH HISTORICAL PERSPECTIVE

Introduction

Immunity is the body's ability to resist or eliminate potentially harmful foreign materials or abnormal cells.

Historical Events in Immunology

Emil von Behring and Kitasato (1890): Serum from vaccinated animals was protective (Diptheria).

Metchnikoff (1880): Cell based Immunity and

Merrill Chase (1940)- Transfer of WBC (immunity to tuberculosis).

1796	**Edward Jenner (smallpox)**
1881	Loius Pasteur (vaccines)
1884	Elie Metchnikoff (phagocytes)
1890	Emil von Behring* (antibodies)
1895	Jules Bordet* (complement)
1906	August Wasserman (syphilis)
1959	Rodney Porter Gerald Edelman* (antibodies)
1960	F McFarlane Burnet* (tolerance)
1975	Cesar Milstein*(monoclonal Ab)
1987	Susumu Tonegawa* (genetics)
1996	Peter Doherty

Immunity

Immunity (immunis- Latin-exempt, state of protection from infectious diseases). Immunity is the body's ability to resist or eliminate potentially harmful foreign materials or abnormal cells consists of the following activities such as defense against invading pathogens (viruses & bacteria), removal of 'worn-out' cells (e.g., old RBCs) & tissue debris (e.g., from injury or disease), Identification & destruction of abnormal or mutant cells (primary defense against cancer), Rejection of 'foreign' cells (e.g., organ transplant), **Inappropriate responses:** Allergies - response to normally harmless substances and autoimmune diseases.

Immune System

- Innate (Nonspecific): 1° line of defense (Pre-existing and Non-specific).
- Adaptive (Specific) 2° line of defense (Responsive and Specific).

Innate Immunity Vs. Adaptive Immunity

Cells of the immune system

Phagocytic cells

Nonspecific

Specific

Lymphocytes

B lymphocytes

T lymphocytes

T helper

T cytotoxic

Two Arms of the Adaptive Immune System

- Humoral immunity
- Antibodies
- Cell-mediated immunity
- T-cells

Immunological Memory

It is more rapid, intense and prolonged.

Chemical Mediators

Extracellular

- Cytokines
- Complement

Intracellular

- Antigens and Antibodies

Antigens

- Chemical makeup
- Types of antigens

Antibodies

- Chemical makeup
- Roles

Innate Immune System: Components of Blood

The innate immune system is made of defences against infection that can be activated immediately once a pathogen attacks. The innate immune system is essentially made up of barriers that aim to keep viruses, bacteria, parasites, and other foreign particles out of your body or limit their ability to spread and move throughout the body. The innate immune system includes:

Physical Barriers: such as skin, the gastrointestinal tract, the respiratory tract, the nasopharynx, cilia, eyelashes and other body hair.

Defence Mechanisms: such as secretions, mucous, bile, gastric acid, saliva, tears, and sweat.

General Immune Responses: such as inflammation, complement, and non-specific cellular responses. The inflammatory response actively brings immune cells to the site of an infection by increasing blood flow to the area. Complement is an immune response that marks pathogens for destruction and makes holes in the cell membrane of the pathogen (Khan Academy).

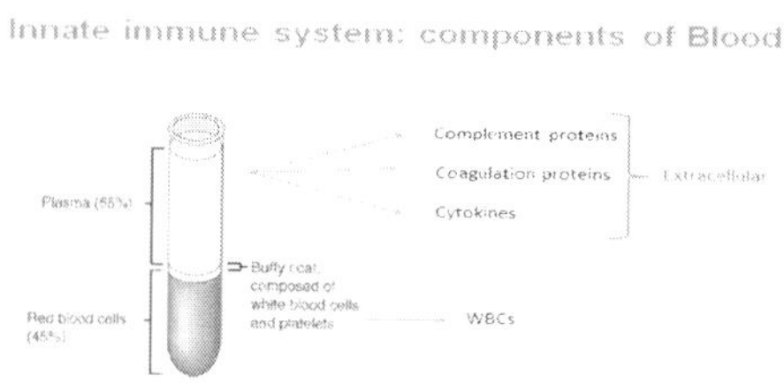

White Blood Cells (WBCS)

White blood cells are part of the immune system and function in immune response.

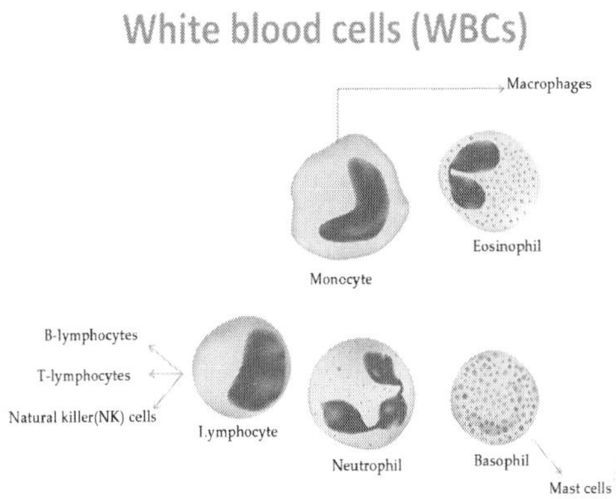

Neutrophils in innate immune response

- Most abundant WBCs (~50-60%)
- Efficient phagocytes
- Most important cells of the innate immune system

Phagocytosis

- Phago = to eat
- Cyte = cell
- WBCs (eg. Neutrophils) – find, eat and digest microbes.

Monocytes

- Monocytes (~5% of WBCs).
- Migrate into the tissues and become macrophages.

Macrophages ("Big eaters")

- Phagocytosis of microbes in tissue (neutrophils are present only in blood).
- Antigen presentation.

Natural killer cells

- It is not B-lymphocytes / T-lymphocytes.
- An important part of the innate immune system.
- Kill virus /bacteria-infected cells (Intracellular pathogens).
- Kills cancer cells.

Toll-like receptors (TLRs)

- Transmembrane proteins.
- Present on macrophages / few other cells.
- Conserved across vertebrates.
- An important part of the innate immune system.
- They look out for microbes (or their components).
- They bind to the microbes (or their components).
- They trigger a cascade of events to kill or protect against pathogens.
- They are innate immune sensors.

Innate response – internal defenses – cellular (WBCs)

- Come into play when the external defenses are breached.
- Neutrophils.
- Monocytes /macrophages.
- NK cells.
- TLRs.

Cytokines

- Small proteins secreted by cells of the immune system.
- Affect the behavior of other cells signaling molecules.
- Key players in innate and acquired immunity.

Cytokines

- Cells of the immune system.
- Neutrophils – when they encounter a pathogen.
- Macrophages – when they encounter a pathogen.
- TLRs – bind to microbe / components of a microbe.
- NK cells – on encountering a microbe infected cell /tumor cell.
- Lymphocytes – when they are activated.

Cytokines

- Interferons

- Interleukins
- Tumor necrosis factor (TNF)

Interferons (IFN)

- The signaling proteins produced by by virus infected monocytes and lymphocytes.
- It is secreted proteins – Key anti-viral proteins.
- "Interfere" with virus replication.
- It is warning the neighboring cells that a virus is around.
- If we did not have IFNs – most of us may die of influenza virus infection.

Interleukins

- Interleukins – 1-37.
- It is not stored inside cells.
- It is quickly synthesized and secreted in response to infection.
- The key modulators of behavior of immune cells.
- It is mostly secreted by T-lymphocytes & macrophages.

Complement

- A large number of distinct plasma proteins that react with one another (C1 thro' C9).
- The complement can bind to microbes and coat the microbes.
- It is an essential part of the innate immune response.
- It enhances adaptive immune response (taught later).

Coagulation proteins

- The coagulation: a mechanism to stop bleeding after injury to blood vessels.
- The complex pathway involves.
- Platelets.
- The coagulation factors.
- Vitamin K

Internal Defenses

Cellular

- Neutrophils
- Monocytes /macrophages
- NK cells
- TLRs

Extracellular

- Cytokines
- Complement
- Coagulation

31. MOLECULAR AND CELLULAR BASIS OF IMMUNE SYSTEM, SELF VERSES NON-SELF RECOGNITION; ANTIGEN – ANTIBODY AND THEIR STRUCTURE, AND FUNCTION

Inflammation

The complex biological process by which the body responds to pathogens and irritants. Associated with swelling of tissue. The key player in the innate immune response fig. 31.1.

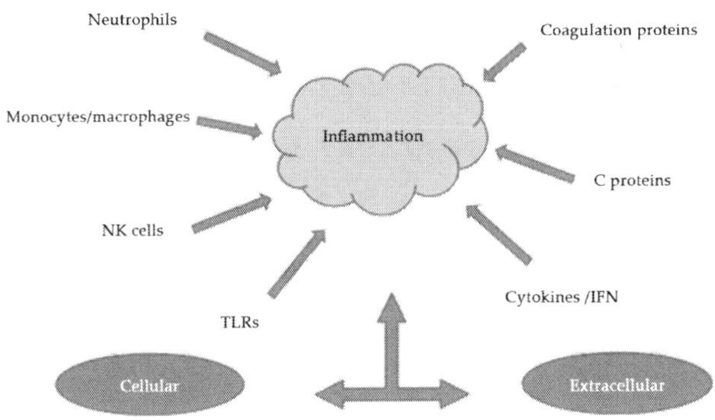

Figure 31.1. Inflammation.

Role of Inflammation in Innate Immunity

- It is an initiation of phagocytosis – killing of pathogens.
- It is limiting the spread of infection.
- It is stimulating an adaptive immune response.
- It initiates tissue repair.

Chronic Inflammation = Tissue Damage

- Chronic inflammation - macrophages in the injured tissue.
- Macrophages release toxins (including reactive oxygen species or ROS) that injure tissues.
- Chronic inflammation is almost always accompanied by tissue destruction.
- Immunogens and antigens

***Immunogen/antigen*:** a substance that elicits an immune response [i.e. a humoral (antibody response) or cell-mediated immune response].

Epitope: the portion of an antigen that is recognized and bound by an antibody (Ab) or a T-cell receptor (TCR) epitope = antigenic determinant fig.31.2

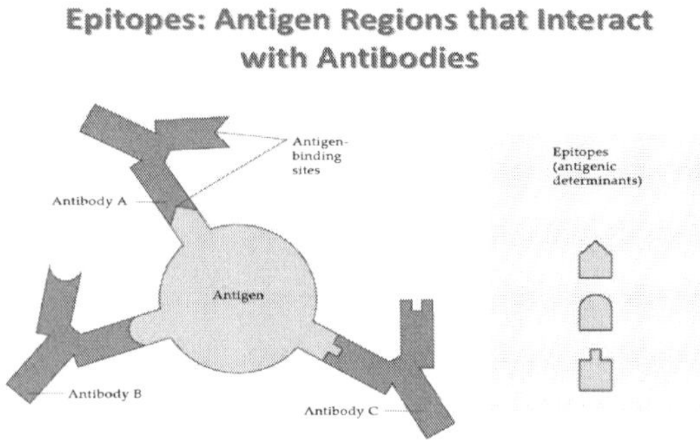

Figure 31.2. Antibody and antigen interaction.

Epitope: the portion of an antigen that is recognized and bound by an Ab or a T Cell receptor. One protein may have multiple antigenic determinants.

B-cell epitopes – recognized by B-cells

T-cell epitopes – recognized by T cells

Immunogenicity

Immunogenicity: is the ability to induce a humoral antibody and/or cell-mediated immune response.

- Weak immunogens
- Strong immunogens

Determines Immunogenicity

Foreignness: It is essential for immunogenicity self-responsive immune cells are eliminated during lymphocyte development.

Chemical composition: Proteins > nucleic acids / polysaccharides / lipids

Structure: Primary /secondary /tertiary structures play a role

Physical form: Particulate> Soluble

Host factors affecting immunogenicity

- Difference across the species (interspecies).
- Differences within a species (intraspecies): Responders/non-responders to vaccine and differences in disease severity in epidemics.

Isoantigens

- **Isoantigens:** Antigens present in some but not all members of a species.
- Blood group antigens – basis of blood grouping.
- MHC (major histocompatibility complex)- cell surface glycoproteins

Autoantigens

Autoantigens are substances capable of immunizing the host from which they are obtained. Self antigens are ordinarily non-antigenic. Modifications of self-antigens are capable of eliciting an immune response.

Haptens are small molecules which are non-immunogenic, thus could never induce an immune response by themselves.

Antibody

It is produced by plasma cells (B-lymphocytes producing Ab). It is an essential part of adaptive immunity. It specifically binds a unique antigenic epitope (also called an antigenic determinant). It possesses antigen-binding sites. The members of the class of proteins are called immunoglobulins fig. 31.3

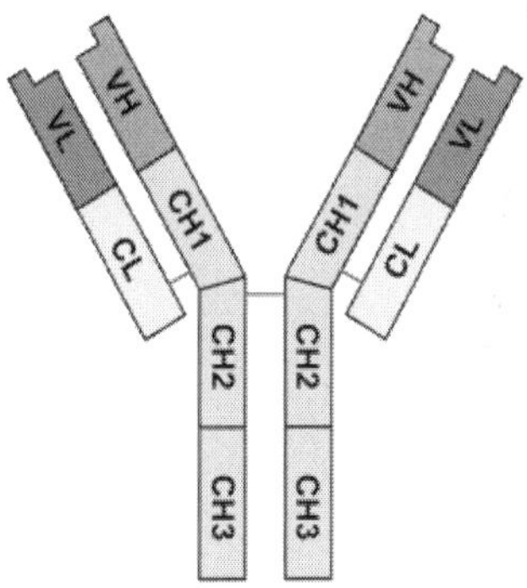

Figure 31.3. Immunoglobulins.

It has 2 identical heavy chains and 2 identical light chains. Each heavy chain has a constant and a variable region. Each light chain has a constant and a variable region fig. 31.4.

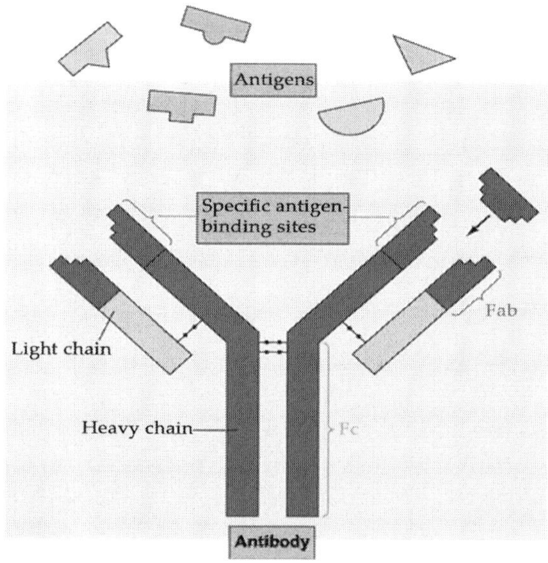

Figure 31.4. Antibody: structure and function.

Antibodies can also be divided into two regions based on their functions. The tip of the "Y", consisting of one constant and one variable domain from each heavy and light chain of the antibody. It contains the region that binds the antigen. It is for this reason that we call it the Fab region (fragment and antigen-binding). The base of the "Y", composed of two heavy chains, plays a role in modulating immune cell activity. This region is called the Fc (Fragment, crystallizable) region. The Fc region can bind to cell receptors, complement proteins, as well as to other immune molecules, thereby initiating different physiological effects including opsonization, cell lysis, and degranulation.

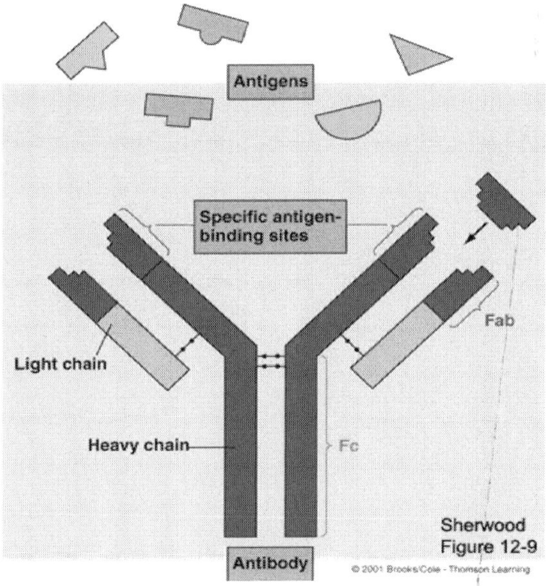

Fab region

It has variable regions. Tip of the antibody and it binds the antigen. The specificity of antigen binding is determined by V_H and V_L.

Antibody: Fc

It is the constant region and base of the antibody and it can bind cell receptors and complement proteins.

Antibodies exist in two forms

Antibodies occur in 2 forms soluble Ag: secreted in blood and tissue and membrane-bound Ag: found on the surface of B-cell, also known as a B-cell receptor (BCR).

Antigen-presenting cells

The process of antigen and "present" it to T cells, dendritic cells, phagocytic and motile. The resident in epithelial surfaces and lymphoid tissues and mature in response to molecular patterns.

Organs of the immune system

Primary immune organs: immune cells develop and mature. Bone marrow:

Differentiation, B cell maturation. Thymus: T cell maturation and Bursa of Fabricius: B cell maturation.

Secondary immune organs: immune cells become activated such as spleen, Lymph nodes and Mucosa-associated lymphoid tissue.

Communicate immune cells

- Soluble mediators
- Extracellular
- Receptor/ligand-specific
- Paracrine vs endocrine
- Signal transduction
- Direct contact:
- Antigen presentation
- Cytotoxicity

Soluble Mediators of the Immune Response

Signaling molecules

Innate: Cytokines, complement

Adaptive: Cytokines, antibodies, chemokines, interleukins

Cytotoxic molecules

Innate: Enzymes, reactive oxygen mediators

Adaptive: Enzymes, reactive oxygen mediators

Cytokines

It has low molecular weight proteins that stimulate or inhibit the differentiation, proliferation, or function of cells. It is multifunctional, overlapping functions. It is produced by immune and non-immune cells. One cell may produce many cytokines. The interleukins: a subset of cytokines that communicate between leukocytes and named consecutively by il- designation.

Chemokines

A subset of cytokines (chemotactic cytokines). A family of **structurally related** cytokines that selectively induce chemotaxis and activation of leukocytes". The old nomenclature: by function mostly. It has new nomenclature: according to N-terminal amino acid sequence such as

CXC= one amino acid separating two N-terminal cysteine residues

CX3C = three amino acids

CC= cysteines adjacent

C= only one cysteine present

32. CONCEPTS OF AG-AB REACTIONS. IMMUNOGLOBULINS - STRUCTURE, BIOLOGICAL CHARACTERS AND FUNCTIONS, ISOTYPES, STRUCTURE AND FUNCTION

Antibody

Antibodies are made up of: 2 light chains (identical) ~25 kDa and 2 heavy chains (identical) ~50 kDa. each light chain is bound to a heavy chain by disulfide (Heavy and light). Heavy chain bound to heavy chain (Heavy-Heavy). It has first 100 a/an of amino terminal vary of both heavy and light chain are variable. It has referred to as v_l, v_h, c_h and c_l. Complementarity (complementarity determining regions) is what binds Ag and the remaining regions are very similar within the same class.

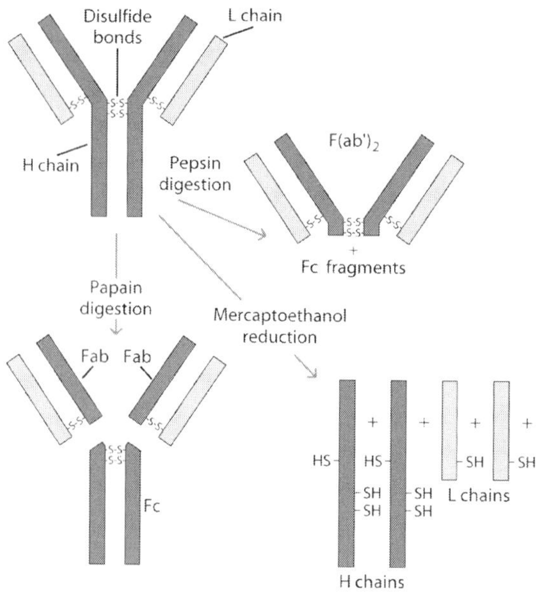

Repeating domains of ~110 a/a, intrachain disulfide bonds within each domain. Heavy chains. 1 V_H and either 3 or 4 C_H (C_H1, C_H2, C_H3, C_H4). Light chains: 1 V_L and 1 C_L.

Hinge Region

It is rich in proline residues (flexible). Hinge is found in IgG, IgA and IgD. It has proline residues that are target for proteolytic digestion (papain and pepsin). It is rich in cysteine residues (disulfide bonds) and IgM and IgE lack hinge regions. They instead have an extra C_H4 Domain.

Enzymatic Digestion of Antibodies

The digestion with papain yields. It has 3 fragments and 2 identical Fab and 1 Fc. Fab because fragment that is antigen binding. Fc because found to crystallize in cold storage. It has pepsin digestion F(ab`)2. No Fc recovery, digested entirely and mercaptoethanol reduction (eliminates disulfide bonds) and alkylation showed.

Sequencing of Heavy Chains

sequencing of several immunoglobulins revealed 100-110 amino terminus, highly variable (V). Five basic sequence patterns α,γ, δ, ε, μ (IgA, IgG, IgD, IgE and IgM). The above classes are called isotypes. Each class can have either κ or λ light chains. Minor differences led to sub-classes for IgA and IgG, IgA1, IgGA2 and IgG1, IgG2, IgG3, IgG4.

Monoclonal Antibodies

Immunize an animal with an antigen and has multiple clones are generated, good for in vivo for clinical diagnosis, research, one clone that reacts to single epitope is preferred. The solution by Kohler and Milstein fuse a myeloma cell (cancerous) with a normal plasma cell. The resulting clones can be cultured indefinitely and produce an antibody recognizing one epitope.

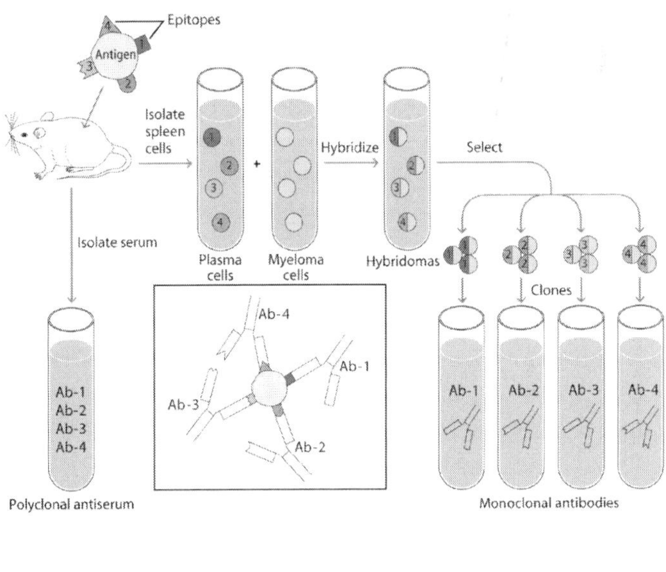

Monoclonal Antibody Applications

Diagnostic Tests

Antibodies are capable to detect tiny amounts (pg/mL) of molecules eg. pregnancy hormones and diagnostic imaging: mAbs that recognize tumor antigens are radiolabeled with iodine I-131. Immunotoxins: mAbs conjugated with toxins and mAbs to clear pathogens.

B-cell Receptor

BCR is an antibody on surface of cell mIg. It has very short cytoplasmic tail and cannot transduce signal heterodimeric molecule ig-α/ig-β transduces (long cytoplasmic tail).

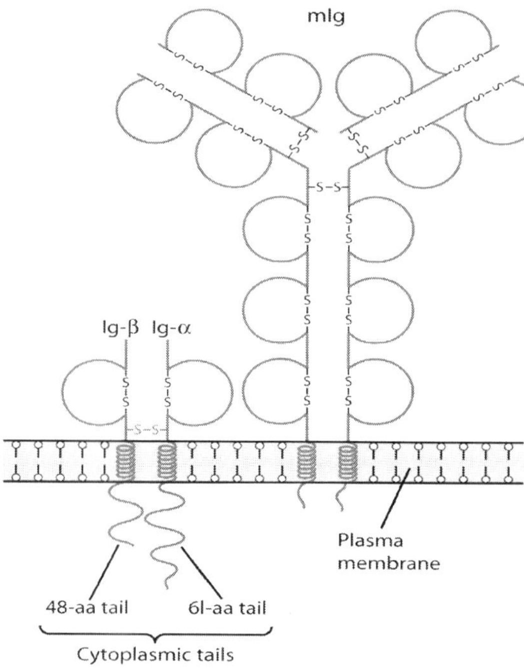

Antibody Classes and Biological Activities

IgG

- Most abundant immunoglobin: 80% of serum Ig~10mg/mL.
- IgG1,2,3,4 (decreasing serum concentration).

- IgG1, IgG3 and IgG4 cross the placenta.
- IgG3 Most effective complement activator.
- IgG1 and IgG3 High affinity for FcR on phagocytic cells and good for opsonization.

IgM

- Five and ten percent of serum immunoglobulin: 1.5mg/Ml.
- mIgM (also IgD) is expressed on B-cells as BCR.
- Its pentameric version is secreted.
- First Ig of the primary immune response.
- It has high valence: Ig (10 theoretical) and 5 empirical.
- It is more efficient than IgG in complement activation.

IgA

- Ten to 15 percent of serum Ig.
- It is the predominant Ig in secretions.
- It is present in milk, saliva, tears and mucus.
- 5-15 g of IgA is released in secretions.
- The serum mainly contains monomeric and polymers.
- The secretions are as dimer or tetramer+J-chain polyptetide+secretory component (Poly IgR).

IgE

- It has very low serum concentration: 0.3μg/mL.
- It participates in immediate hypersensitivities reactions eg. asthma, anaphylaxis, hives
- It binds mast cells and blood basophils.
- Its binding causes degranulation (Histamine Release).

IgD

- It is expressed on B-cell surface.
- IgM and IgD areexpressed on b-cell surface.

Antibodies Act As Immunogens

Antigenic determinants on Abs fall in three categories.

- Isotypic
- Allotypic
- Idiotypic

Isotypic

- It has the constant region of Ab.
- If you inject Ab in a different species- anti-isotype is generated.
- If it is within the same species there is no anti-isotype.

Allotype

Even though the same isotypes occur within one species, small differences (1-4 a/a) arise in different individuals (a form of polymorphism). If injected with such Ab you generate anti-allotype Ab eg. during pregnancy and blood transfusion.

Idiotype

Unique V_H and V_L binds antigen but can also behave as antigenic determinants. If you inject a monoclonal antibody into a genetically identical recipient, then anti-idiotypic antibodies are generated. No anti-isotypic and no anti-allotypic Abs will be generated.

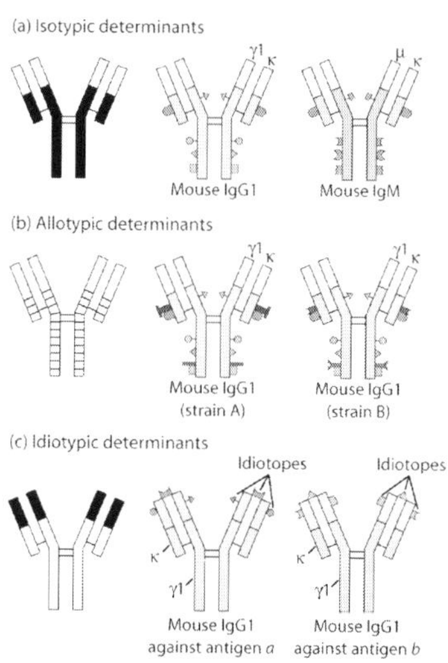

Antigen Recognition

B cells can recognize antigens via their surface Ig molecules. T cells can only recognize antigen in association with a Major Histocompatibility Complex (MHC) molecule.

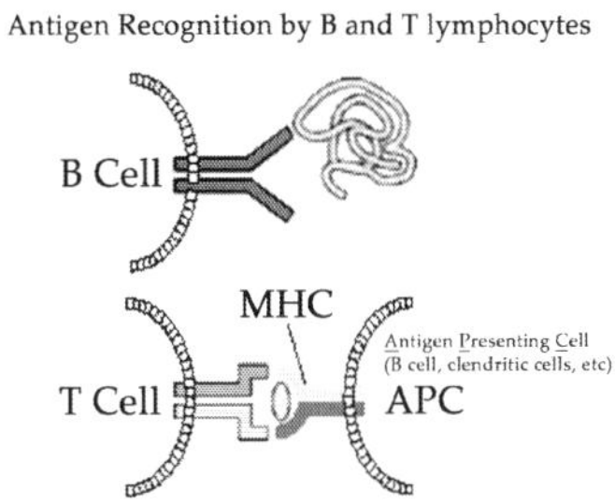

Immunoglobin Fold

V and C domains share the basic Ig fold. It has differences between the two domains. C domain is built of seven b-strands arranged so that four strands form one sheet and three strands form a second sheet. The strands are closely packed together and joined by a single disulfide bond. It has three most of the invariant residues of the constant domain are in the sheets. The overall structure of the V domain is very similar but there are nine strands instead of seven. The two additional strands harbor CDR2.

Complementarity Determining Regions

Six loops of the VH (H1, H2 and H3) and VL (L1, L2 and L3) domains create a great variety of surfaces. It has deep binding cavities: such as those seen in some antibody-hapten complexes. It has wide pockets: seen in certain antibody-peptide complexes. It has flat surfaces: seen in antibody-protein interactions and H3 is the most variable of the loops and in all crystallographically solved antibody-antigen complexes makes several contacts with antigen.

Antigen: Antibody Complex

Antibodies bind to antigens by recognizing a large surface, and through surface complementarity. Thus, these complexes have a very high affinity for each other.

Weak Forces Vs High Affinity

The interaction between an antigen and antibody can be very strong, and yet all of the forces involved are considered to be relatively weak. How can weak hydrogen bonds, electrostatic attractions, hydrophobic forces, and van der Waals contacts lead to a high affinity. Its contact between antigen and antibody occurs over a wide surface area, allowing multiple weak interactions that give a strong affinity. The hydrogen bonds join the antibody and antigen over a wide surface area. Other weak forces, including van der Waals forces, electrostatic attractions and hydrophobic forces, add to the strength and specificity of antibody/antigen binding.

Antibody-Hapten Complex

Haptens, having a limited total surface area, deeply embed themselves into the VL/VH dimer interface. Hapten binding antibodies frequently show a deep central cavity, long CDR L1 loops and a CDR H3 loop with an "open" conformation, allowing the hapten to bind as much as 80% of its total surface in the interaction.

Protein Antibody Complex

In contrast, proteins preferentially to a relatively flat binding surface. In a "closed" CDR H3conformation, the CDR H3 loop packs down onto the central cavity, and the protein antigen binds on top of it.

Effector Response is Mediated Via Ig-Fcr Complex Formation

Antibodies not only must recognize an antigen, but also must invoke responses – effector functions – that will remove the antigen and kill the pathogen. The variable regions of antibody are the sole agents of binding to antigen. The heavy chain constant region (C_H) is responsible for interactions with other proteins (e.g. complement), cells (elements of the innate immune system), and tissues that result in the effector functions of the humoral response. FcR recognizes the Fc portion of antibodies not antigens.

The FC-FC Receptor Complex

FcR plays important role in antibody-mediated immune responses. Ig and FcR binding activates effector functions. Fc receptor interacts with the CH2 and CH3 domains of immunoglobulins.

Immune Recognition: MHC And TCR Interactions

T cells can only recognize antigen in association with a major histocompatibility complex (MHC) molecule.

T cells

T cells display TCR as their antigen recognition protein. When stimulated they become Cytotoxic.

Helper T cells

Secrete cytokines that recruit other cells of the IS. TCR's only recognize short peptides.

MHC & T cells

T cells have a requirement to recognize both the antigen and the MHC molecule. This is because the molecular structure of the MHC-antigen complex is arranged so that some of the polymorphic amino acids of the MHC molecule are in direct contact with the TCR Therefore T cell recognition of antigen is said to be MHC 'restricted'.

Antigen processing and presentation

There is the fragmentation of protein into peptides and association of peptide with an MHC molecule and transport to the cell surface for expression. It has different cellular pathways for the association of peptides with MHC class I and class II molecules.

MHC & Antigens

MHC Class I

It presents endogenously derived peptides. These can be either self or derived from viruses because MHC Class I is present on all cells any cell can interact with T cells if infected by a virus. It is detected on all nucleated cells. It has very highly polymorphic. It has tight fit for peptides of only about 9 aa. It has consisted of an a-chain of 3 domains associated with b-2 microglobulin.

MHC Class II

It presents exogenous antigens that has been phagocytosed and processed eg. bacteria. This is performed by professional antigen presenting cells eg. macrophages. It is seen only on the 'professional antigen processing cells' e.g macrophage slightly less polymorphic accepts peptides of up to fifteen amino acids.

MHC Class I And Class II Proteins

Class I

Alpha chain

It has three external domains, one transmembrane, one cytoplasmic tail and encoded in MHC.

Beta-2 microglobulin

It has one external domain, no transmembrane, no cytoplasmic tail and not encoded in MHC.

Class II

Alpha chain

It has two external domains, one transmembrane, one cytoplasmic tail and encoded in MHC.

Beta chain

It has two external domains, one transmembrane, one cytoplasmic tail and encoded in MHC.

Peptide Binding By Major Histocompatibility Complex (MHC) Antigen-Presenting Proteins

MHC I

It has peptides of intracellular origin. It has peptides nine-ten residues long. It has deep pockets that bind peptide sidechains. It has dccp pockets that bind peptide N- and C-termini.

MHC II

It has peptides of extracellular origin. It has peptides fifteen residues or longer. It has shallow pockets that bind peptide sidechains and termini-free. It has H-bonds to the peptide backbone.

MHC Polymorphism

It has both classes I and class II genes: are highly polymorphic. It has most polymorphic residues of class I are in the alpha 1 and alpha 2 domains. It has most polymorphic residues of Class II are in the alpha 1 and beta 1 domains.

33. THEORIES OF ANTIBODY SYNTHESIS, COMPLEMENT SYSTEM AND THEIR REACTIONS. ORGANISMS AND CELLS INVOLVED IN IMMUNITY

Antibodies are protein molecules secreted by immune system cells that help the body fight invaders. Each antibody-producing cell makes a single, unique antibody which will be in one of the 5 major classes – IgA, IgD, IgE, IgG, and IgM. These classes are determined by the constant region of heavy chains of the antibody, and have different immunologic targets and mechanisms of action. They can also bind their targets with different degrees of strength, depending on the fit between the antibody and its target (Khan Academy).

Nature of Ag/Ab Reactions: It has the lock and key concept

- Non-covalent bonds
- Hydrogen bonds
- Electrostatic bonds
- Van der Waal forces
- Hydrophobic bonds
- Multiple bonds
- Reversible

Strength of the reaction between a single antigenic determinant and a single Ab combining site.

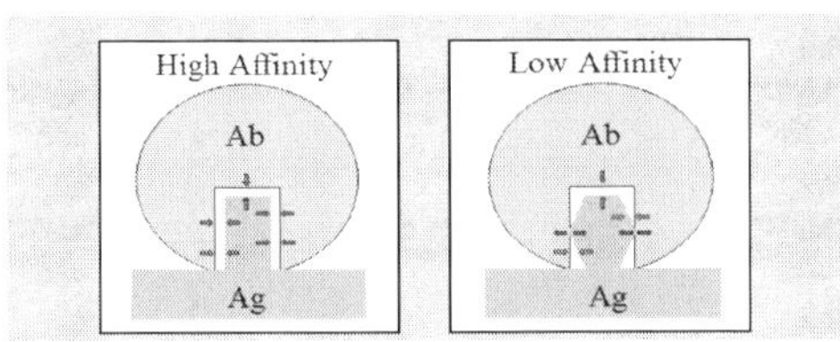

Affinity = ∑ attractive and repulsive forces

Calculation of Affinity

Ag + Ab   Ag-Ab

Applying the Law of Mass Action

Applying the Law of Mass Action:

$$K_{eq} = \frac{[Ag\text{-}Ab]}{[Ag] \times [Ab]}$$

Avidity

The overall strength of binding between an Ag with many determinants and multivalent Abs specificity. The ability of an individual antibody combining site to react with only one antigenic determinant. The ability of a population of antibody molecules to react with only one antigen.

Cross-Reactivity

The ability of an individual Ab combining site to react with more than one antigenic determinant. The ability of a population of Ab molecules to react with more than one Ag.

Factors Affecting the Measurement Of AG/AB Reactions

- Affinity.
- Avidity.
- Ag: Ab ratio.
- It has the physical form of Ag.

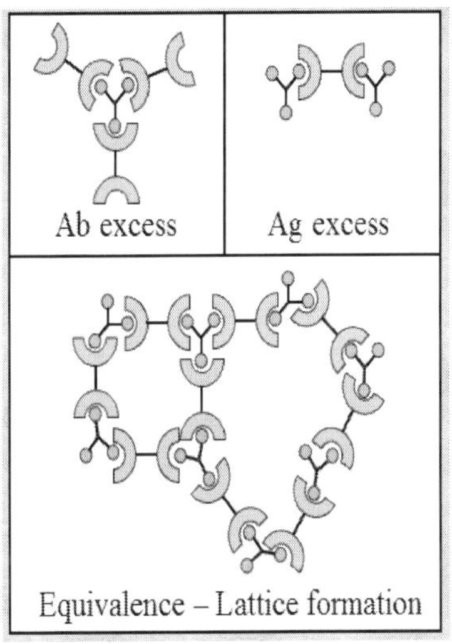

Tests Based On AG/AB Reactions

All tests based on Ag/Ab reactions will have to depend on lattice formation or they will have to utilize ways to detect small immune complexes. All tests based on Ag/Ab reactions can be used to detect either Ag or Ab

Agglutination/Hemagglutination

Definition - tests that have as their endpoint the agglutination of a particulate antigen: Agglutinin/hemagglutinin

Qualitative agglutination test: Ag or Ab

Quantitative agglutination test: Titer and Prozone

Precipitation Tests

- Radial immunodiffusion (Mancini)
- Immunoelectrophoresis
- Radioimmuoassays (RIA)
- Enzyme-Linked Immunosorbent Assays (ELISA)

Cells and Organs of the Immune System

Cells – leukocytes

- Lymphocytes – specialized for adaptive immunity.
- Granulocytes/Agranulocytes – function in accessory roles in adaptive immunity.

Immune Organs – 2 major groups

- 1° lymphoid organs – where lymphocytes develop and mature.
- 2° lymphoid organs – where lymphocytes interact w/ Ag.

Hematopoiesis

It begins with hematopoietic stem cells (HSC): It is less in the bone marrow; difficult to culture and pluripotent able to produce RBC's, WBC's, megakaryocytes, HSC differentiates to become either a) myeloid progenitor cell b) Lymphoid progenitor cell.

Myeloid →RBC's and WBC's and dendritic cells

Lymphoid →B and T cells and dendritic cells

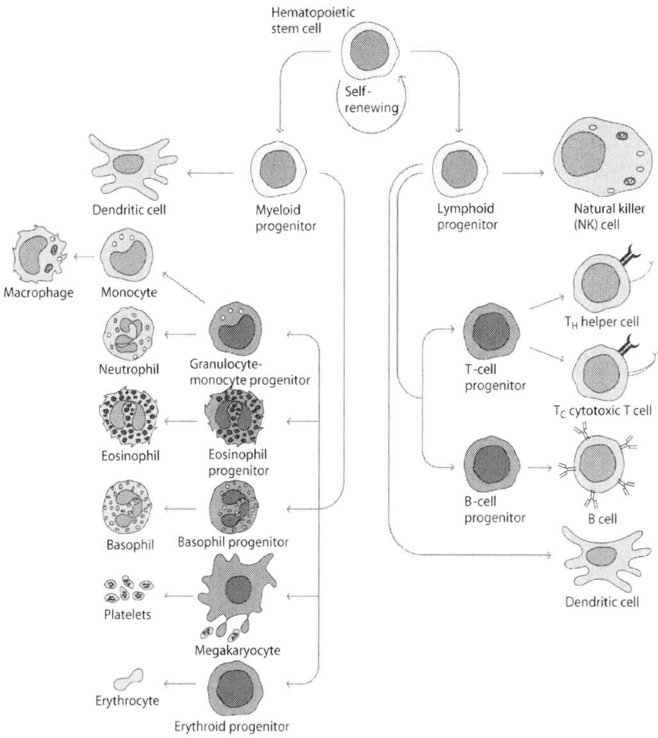

Hematopoietic Influences

Stromal cells influence diff of HSC by providing hematopoietic-inducing microenvironment (HIM) with colony-stimulating factor (CSF) and -erythropoietin (EPO).

Cytokine influences > stimulated by activated macrophages and T cells

Genetic Regulation

Hematopoietic homeostasis is required to maintain certain cell balance between cells removed by cell death and those produced for WBC's: a human must produce ~3.7 x 10^{11}/day. Each WBC has a specific life span + dies by apoptosis; neutrophils live only a few days.

Apoptosis – Programmed Cell Death

Cells die in this process + later phagocytosed, decrease in cytoplasmic volume; apoptotic bodies, clumping/break up of DNA, phagocytosed by macrophages → blocks the release of cell contents no local inflammatory response. It is influenced by gene expression and chief among the genes is bcl2 - which inhibits cell death.

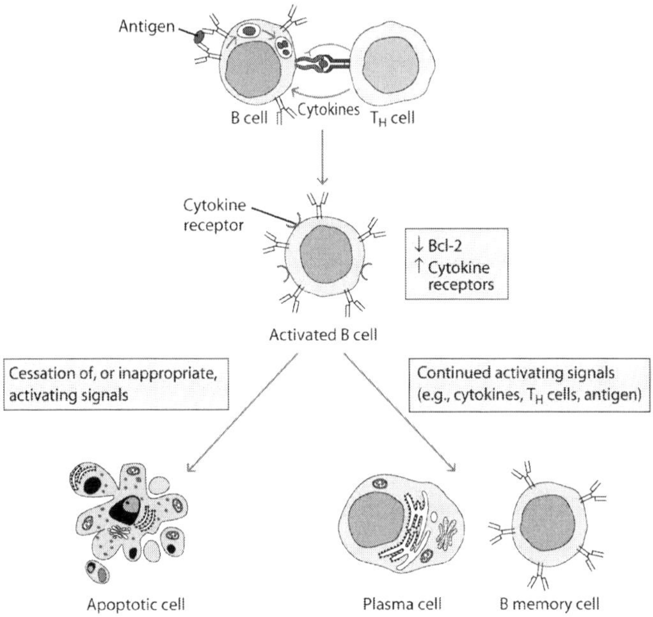

34. THE HUMORAL IMMUNE RESPONSE; ANTI-BODIES-STRUCTURE, DIVERSITY OF FUNCTION AND MECHANISM OF ACTION

Pathogens, agents that cause disease, infect a wide range of animals, including humans. The immune system recognizes "foreign bodies and responds with the production of immune cells and proteins. All animals have innate immunity, a defense active immediately upon infection. Vertebrates also have adaptive immunity. Innate immunity is present before any exposure to pathogens and is effective from the time of birth. It involves nonspecific responses to pathogens. Innate immunity consists of external barriers plus internal cellular and chemical defenses. Adaptive immunity, or acquired immunity, develops after exposure to agents such as microbes, toxins, or other foreign substances.

It involves a very specific response to pathogens. The immune system recognizes bacteria and fungi by structures on their cell walls. An immune response varies with the class of pathogen encountered. The immune system of mammals is the best understood of the vertebrates. Innate defenses include barrier defenses, phagocytosis, antimicrobial peptides. Additional defenses are unique to vertebrates: natural killer cells, interferons, and the inflammatory response. The barrier defenses include the skin and mucous membranes of the respiratory, urinary, and reproductive tracts. Mucus traps and allows for the removal of microbes. Many body fluids including saliva, mucus, and tears are hostile to many microbes. The low pH of the skin and the digestive system prevents the growth of many bacteria. Pathogens entering the mammalian body are subject to phagocytosis. Phagocytic cells recognize groups of pathogens by TLRs, Toll-like receptors. A white blood cell engulfs a microbe, then fuses with a lysosome to destroy the microbe. There are different types of phagocytic cells

- **Neutrophils** engulf and destroy pathogens
- **Macrophages** are found throughout the body
- **Dendritic cells** stimulate the development of adaptive immunity

Eosinophils discharge destructive enzymes cellular innate defenses in vertebrates also involve natural killer cells. These circulate through the body and detect abnormal cells. They release chemicals leading to cell death, inhibiting the spread of virally infected or cancerous cells. Many cellular innate defenses involve the lymphatic system. The adaptive response relies on two types of lymphocytes, or white blood cells. Lymphocytes that mature in the thymus above the heart are called T cells, and those that mature in the bone marrow are called B cells. Antigens are substances that can elicit a response from a B or T cell. Exposure to the pathogen activates B and T cells with antigen receptors specific for parts of that pathogen. The small accessible part of an antigen that binds to an antigen receptor is called an epitope. B cells and T cells have receptor proteins that can bind to foreign molecules. Each individual lymphocyte is specialized to recognize a specific type of molecule.

Antigen Recognition by B Cells and Antibodies

Each B cell antigen receptor is a Y-shaped molecule with two identical heavy chains and two identical light chains. The constant regions of the chains vary little among B cells, whereas the variable regions differ greatly. The variable regions provide antigen specificity. The binding of a B cell antigen receptor to an antigen is an early step in B cell activation. This gives rise to cells that secrete a soluble form of the protein called an antibody or immunoglobulin (Ig). Secreted antibodies are similar to B cell receptors but lack transmembrane regions that anchor receptors in the plasma membrane. Immunological memory is responsible for long-term protections against diseases, due to either a prior infection or vaccination. The first exposure to a specific antigen represents the primary immune response. During this time, selected B and T cells to give rise to their effector forms. In the secondary immune response, memory cells facilitate a faster, more efficient response. Acquired immunity has two branches: the humoral immune response and the cell-mediated immune response. In the humoral immune response, antibodies help neutralize or eliminate toxins and pathogens in the blood and lymph.

In the cell-mediated immune response, specialized T cells destroy affected host cells. Antibodies do not kill pathogens; instead, they mark pathogens for destruction. In neutralization, antibodies bind to viral surface proteins preventing infection of a host cell. Antibodies may also bind to toxins in body fluids and prevent them from entering body cells. In opsonization, antibodies bind to antigens on bacteria creating a target for macrophages or neutrophils, triggering phagocytosis". Antigen-antibody complexes may bind to a complement protein—which triggers a cascade of complement protein activation. Ultimately a membrane attack complex forms a pore in the membrane of the foreign cell, leading to its lysis. B cells can express five different forms (or classes) of immunoglobulin (Ig) with similar antigen-binding specificity but different heavy chain C regions.

- IgD: Membrane-bound
- IgM: First soluble class produced
- IgG: Second soluble class; most abundant
- IgA and IgE: Remaining soluble classes

Effector Functions of Antibodies

Opsonization: is the promotion of phagocytosis of antigens by macrophages and neutrophils. Protein molecules are called Fc receptors (FcR), which bind the constant region of an antibody, are present on the surfaces of phagocytes.

Activation of complement: IgM (most effective) and most subclasses of IgG can activate the classical complement pathway. Antibody-dependent cell-mediated cytotoxicity (ADCC) – The linking of antibody bound to target cells (virus-infected cells, or some tumor cells) with FcR of natural killer cells (NK cells), neutrophils, macrophages, or eosinophils can result in the killing of the target cell. Neutralization of viruses and bacteria can prevent attachment to cell receptors and neutralization of toxins.

35. THE CELL MEDIATED IMMUNE RESPONSE; RECOGNITION OF ANTIGEN BY T CELLS; ANTIGEN PRECIPITATION; THE MAJOR HISTOCOMPATIBILITY PROTEINS; THE ROLE OF CYTOKININS AND THE REGULATION OF IMMUNE RESPONSES, HYPERSENSITIVITY REACTIONS

In the cell-mediated immune response, specialized T cells destroy affected host cells. A type of T cell called a "helper T cell triggers both the humoral and cell-mediated immune responses. Signals from helper T cells initiate the production of antibodies that neutralize pathogens and activate T cells that kill infected cells. Antigen-presenting cells have class I and class II MHC molecules on their surfaces. Class II MHC molecules are the basis upon which antigen-presenting cells are recognized. Antigen receptors on the surface of helper T cells bind to the antigen and the class II MHC molecule; then signals are exchanged between the two cells. The helper T cell is activated, proliferates, and forms a clone of helper T cells, which then activate the appropriate B cells.

Cytotoxic T Cells: A Response to Infected Cells

Cytotoxic T cells are the effector cells in the cell-mediated immune response. Cytotoxic T cells recognize fragments of foreign proteins produced by infected cells and possess an accessory protein that binds to class I MHC molecules. The activated cytotoxic T cell secretes proteins that disrupt the membranes of target cells and trigger apoptosis fig. 35.1.

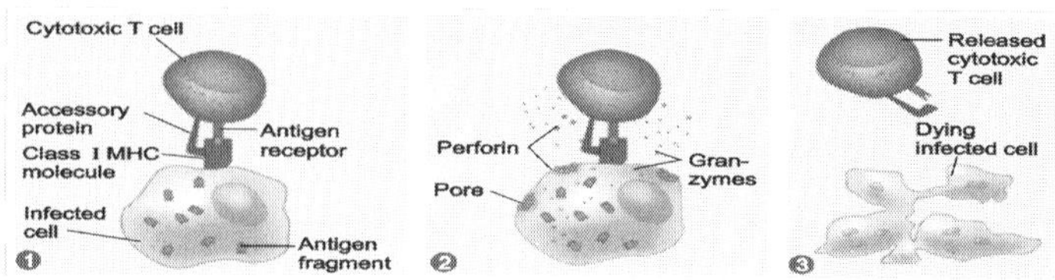

Figure 35.1. Cytotoxic.

Effector T Cells

The membrane-bound molecules eg. FASL (CD8+). The soluble molecules CTLs secrete cytotoxins perforins and granzymes also secret IFN-γ and TNF-β.

Cytotoxic T Cells

CTLs recognize cells that have been infected by some viruses transformed to tumor. CTL activation is divided into 2 phases. Activation and differentiation of naïve CTL. Effector recognizes Class I MHC/peptide and destroys target naïve CTLs cannot kill. It is referred to as CTL-Ps (precursors). It has 3 signals needed for activation. Ag-specific signal via TCR/MHC I+Ag. The co-stimulatory signal CD28(CTL)/B7 (APC). IL-2 signaling induces proliferation (CTL-P do not express IL-2 R). IL-2 is provided by T_{H1} or CTL-P itself. IL-2R is expressed only after activation. T_{H1} and CTL collaborate to induce effector CTL. IL-2 seems to be crucial (knockout data). Lack of IL-2R in CTL-P ensures Ag specificity. Upon clearance of antigen CTLs undergo apoptosis. T_{H1} induces up-regulation of Co-stimulatory molecules on APCs enhancing APC/CTL co-stimulation.

Conjugate Formation

LFA-1 (CTL) binds ICAMs (Target), LFA-1 changes to high avidity if Ag Is Recognized, Activated LFA-1 persists for 5-10 mins, Membrane attack, Requires Ca^{2+} and energy. Granules release Perforins (65 kDa) and Granzymes (serine proteases) at the junctional space. Perforins polymerize forming cylindrical pores (5-20 nm), Ca^{2+} is needed". Granzymes enter the target cell. Granzyme B can enter via mannose-6-phosphate receptor in a vesicle. Lytic enzymes (MΦ, Neutrophils, Eosinophils, NK).

Cytokines

They are low molecular weight proteins, <30kD, high affinity for receptors and are active in picomole amounts fig 35.2.

Chain of Cytokine Action

Stimulus>cytokine-producingcell>cytokine>targetcell>receptor>biological effect(s).

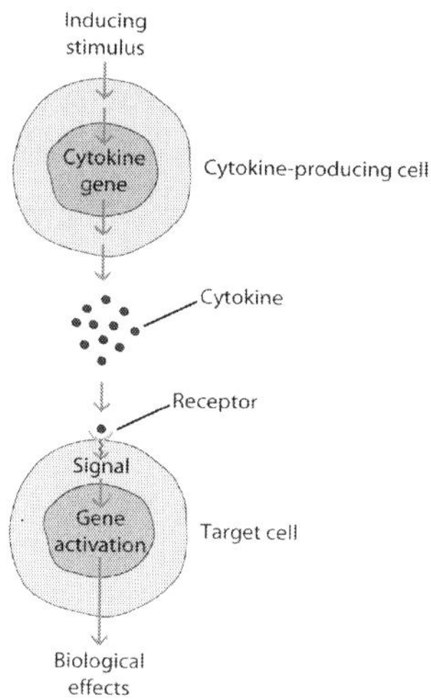

Figure 35.2. Cytokines.

Types of Cytokines

- Lymphokines
- Function e.g., Chemokines

Intercellular action e.g., Interleukins

Action of Cytokines

Autocrine: Affects the generating cell (self)

Paracrine: Affects cells in the immediate vicinity

Endocrine: affects cells remote from the secreting cell.

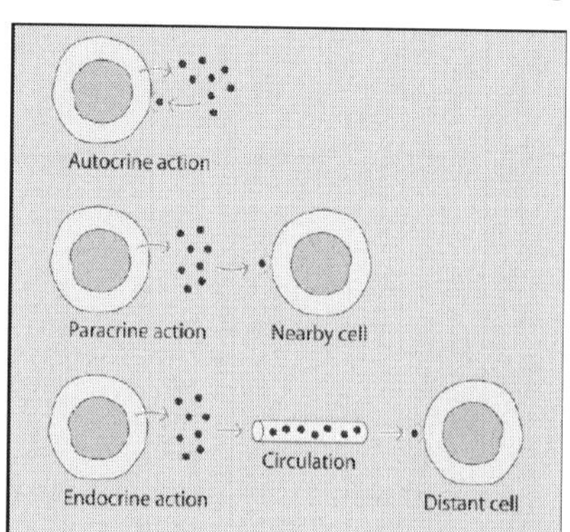

Figure 35.3. Endocrine.

Pleiotrop: affects multiple cell types.

Redundancy: multiple cytokines affect cells of the same type.

Synergy: cytokines acting in concert on the same cell.

Antagonism: competing actions.

Cascading: cytokines acting sequentially fig. 35.4.

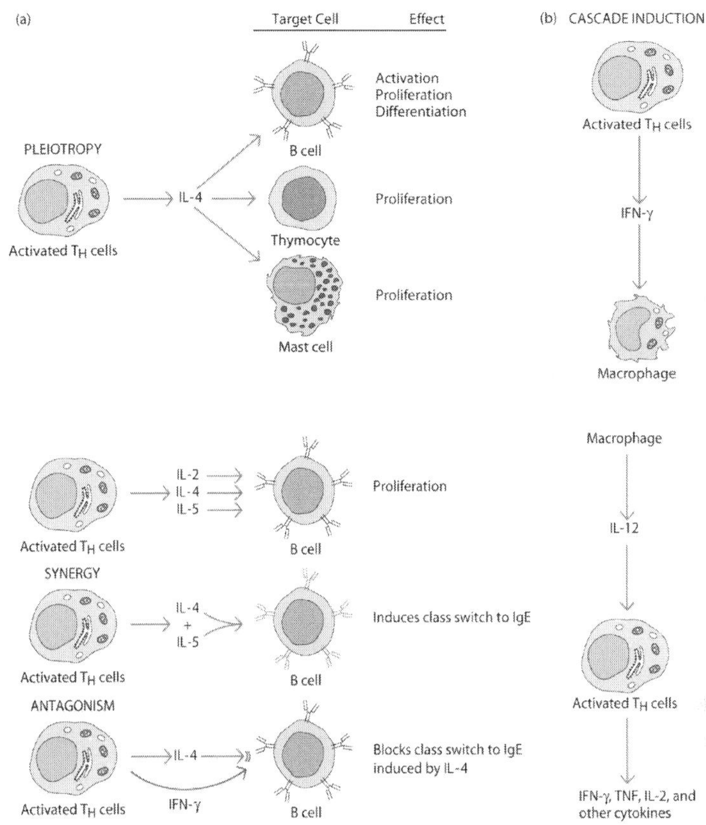

Figure 35.4. Overview of immune responses.

Cytokine-Generating Cells

- Innate immunity
- Macrophages
- Endothelial cells
- Fibroblasts
- Adaptive immunity

- T lymphocytes
- Macrophages
- NK cells

Cytokine Receptors

- Ig super family
- Class I- hematopoietin
- Class II-interferon
- TNF
- Chemokine
- It has multimeric receptors
- It is common signal-transducing subunits
- It has unique high affinity subunits
- It has high affinity subunits associated with activation of the target cell

Signal Transduction

- It is initiated by cytokine binding.
- The activates of JAK (Janus kinase) signaling.
- Phosphorylation of tyrosine process.
- The binding of STAT (signal transducers and activators of transcription) pathway.
- The translocation of STAT to cell's DNA.
- The transcription of specific target genes.

Role of Cytokines In Hematopoiesis

- SCF (Stem cell factor)
- GM-CSF (Granulocyte-macrophage colony-stimulating factor)
- IL-3 (Interleukin 3)
- IL-5 (Interleukin 5)
- IL-7 (Interleukin 7)
- IL-11(Interleukin 11)

Cytokines in the Immune Response

Innate Immune Response

- IL 1-(Macrophage)-fever, capillary effects
- IL 6-(Macrophage)-adaptive immunity via B cells
- IL 12(Macrophage)-adaptive immunity via T helper cells
- TNF (Macrophage)-capillary effects, activates neutrophils
- IFN alpha (Macrophage)-multiple effects
- IFN beta (Fibroblasts)-multiple effects

Adaptive Immune Response

- IL 2- (T cells)-multiple effects).
- IL 4- (T cells & mast cells)-T cell differentiation, IgE production.
- TGF beta – (T cells, macrophages)-inhibits the adaptive immune response.
- IFN gamma- (T cells, NK cells)-Macrophage activation.

Subsets of T Helper Cells

- T helper 1
- IgG production/Complement fixation
- Macrophage activation
- DTH
- CTL production
- T helper 2
- IgE production
- Eosinophils

Major Histocompatibility Complex, AKA MHC

A complex of genes encoding "cell-surface molecules that are required for antigen presentation to T-cell.

The basis of self / not self distinction, presentation of processed antigen. MHC-I (on nearly all nucleated cells)-Class I and MHC-II (on B-cells, macrophages, dendritic cells) Class II.

Major histocompatibility complex

Cluster of genes found in all mammals, its products play role in discriminating self/non-self and participant in both humoral and cell-mediated immunity. MHC act as antigen presenting structures, in humans, MHC is found on chromosome 6 referred to as HLA complex, in mice MHC is found on chromosome 17 referred to as H-2 complex fig 35.5.

MHC-I

Heavy chain (alpha) and "microglobulin" (beta two). Heavy is 45 kilodaltons, has three domains + a transmembrane component (40 aa) + a cytoplasmic tail (30 aa). The three alpha domains are called: α1, α2, & α3, α1 and α2 interact to present processed Ag. The process of Ag is optimally a nonomer, Microglobulin (12 kDa) associates non-covalently with α3, Microglobulin and α3 are part of the immunoglobulin superfamily, microglobulin is the only member of the superfamily that does not have a component linking it to a membrane.

MHC-II

An alpha and beta chain, 33 kDA and 28 kDa, respecitvely. The chains are non-covalently associated. Each chain has two domains like α1-β1 interact to present processed Ag. The processed Ag is optimally 13-18 aaα2 & β2 are part of immunoglobulin superfamily.

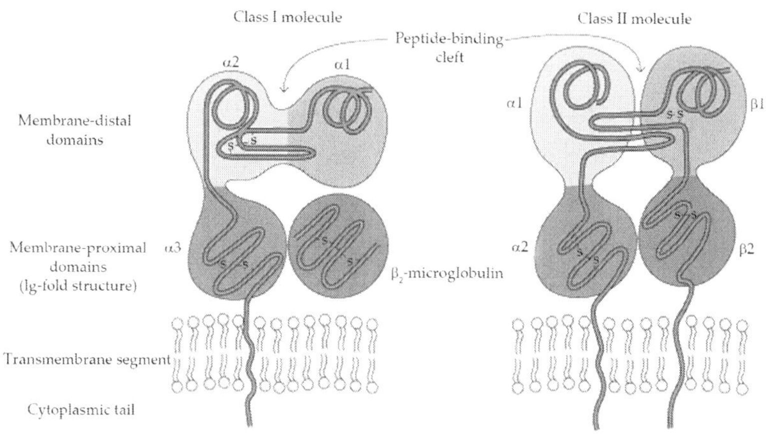

Figure 35.5. Major histocompatibility complex.

Genes of MHC Organized in 3 Classes

Class I MHC genes

- Glycoproteins expressed on all nucleated cells.
- Major function to present processed Ags to T_C.

Class II MHC genes

Glycoproteins expressed on MΦ, B-cells, DCs.

Major function to present processed Ags to $T_{H.}$

Class III MHC genes

The products include secreted proteins that have immune functions eg. complement system, inflammatory molecules. Class I MHC genes found in regions A, B and C in humans (K and D in mice). Class II MHC genes found in regions DR, DP and DQ (IA and IE in mice). Class I and Class II MHC share structural features. Both involved in APC". Class III MHCs have no structural similarity to Class I and II eg. TNF, heat shock proteins and complement components fig. 35.6.

Mouse H-2 complex

Complex	H-2						
MHC class	I	II		III		I	
Region	K	IA	IE	S		D	
Gene products	H-2K	IA αβ	IE αβ	C′ proteins	TNF-α TNF-β	H-2D	H-2L

Human HLA complex

Complex	HLA							
MHC class	II			III		I		
Region	DP	DQ	DR	C4, C2, BF		B	C	A
Gene products	DP αβ	DQ αβ	DR αβ	C′ proteins	TNF-α TNF-β	HLA-B	HLA-C	HLA-A

Figure 35.6. Classes of MHC.

MHC Expression

Expression is regulated by many cytokines: IFNα, IFNβ, IFNγ and TNF increase MHC expression. Transcription factors that increase MHC gene expression: CIITA (transactivator), RFX (transactivator). Some viruses decrease MHC Expression: CMV, HBV, Ad12 reduction of MHC may allow for immune system evasion.

36. LYMPHOCYTES. DIFFERENTIATION OF STEM CELLS AND GENERATION OF LYMPHOCYTES, MECHANISMS

Lymphocytes are derived in bone marrow from stem cells include both T cells and B cells. "T cells: stored & mature in thymus-migrate throughout the body killer cells. It performs the lysis (infected cells), cell-mediated immune response, Helper cells are enhancing T killer or B cell activity. Suppressor cells reduce/suppress immune activity and it may help prevent autoimmune disease.

B-cells are stored and mature in spleen, secrete highly specific Ab to bind foreign substances (antigen: Ag), form the Ab-Ag complex, responsible for humoral response, perform antigen processing and presentation, differentiate into plasma cells (large Ab secretion). Neutrophils are found throughout the body, in blood phagocytosis of Ab-Ag CX. Macrophages are produced throughout the body, blood, lymphatics phagocytose non-specifically (non Ab coated Ag) : Ab-Ag CX. These have a large number of lysosomes (degradative enzymes). It performs Ag processing and presentation. It present Ag to T helper cell. It secretes lymphokines/ cytokines to stimulate T helper.

Natural killer cells- in the blood throughout the body, destroy cancer cells and are stimulated by interferons.

Severe Combined Immunodeficiency Disease (SCID)

It is generally caused by a defect of a single gene needed for T-cell and B-cell function and the subject exhibits no cell-mediated response and cannot make antibodies. In about 25% of cases , the defective gene for the enzyme adenosine deaminase (required for purine breakdown) is involved.

37. IMMUNITY TO INFECTIOUS DISEASES, VACCINES. GENERATION OF MONOSPECIFIC ANTIBODIES: HYBRIDOMA TECHNOLOGY

Introduction

Despite innate and adaptive immune responses to pathogens, infectious diseases have plagued human populations throughout history still cause millions of deaths per year. There are four main types of pathogens that cause infectious disease such as

1. Viruses
2. Bacteria
3. Protozoa
4. Helminths

One of the four main pathogens responsible for infectious diseases which responsible for smallpox, the common cold, chickenpox, influenza, shingles, herpes, polio, rabies, Ebola and AIDS. Several specific immune effectors & nonspecific defense mechanisms. Viruses act to subvert one or more of these mechanisms to prolong their survival.

Viruses: Structure & Function

Viruses depend on host cells for reproduction. The viruses remain metabolically inert outside of host cells. They exist as a protein coat or capsid, sometimes enclosed within a membrane. The capsid encloses either DNA or RNA which codes for the virus elements.

INNATE IMMUNITY CELLS ARE EXPRESS RECEPTORS THAT RECOGNIZE BACTERIAL MOLECULES

- TLR-4 (Recognizes LPS, gram negative bacteria)
- TLR-2 (Recognizes PGN, gram positive bacteria)
- TLR-3,7,9 (viral nucleic acids)

Viruses induce the production of interferons (IFN-α, IFN-β and IFNγ): Interferons produce an anti-viral state. The pathogens use several tricks to avoid the immune system.

Induce a specific type of immunity that allows it to survive. It is a continuous variation of antigens eg. HIV

It expresses antigens that resemble our own genes. Infections are prevalent in developing countries: 4,000,000 deaths due to respiratory infections.

Viral Infections

The immune response against the virus is primarily mediated through interferons. The double-stranded RNA induces the production of IFN; themain producers of IFNα and IFNβ are pDCs. TLR-3 (dsRNA); TLR-7 (ssRNA). Interferons are produced in an anti-viral state: a state that inhibits viral replication.

IFN Signaling

Upon ligand binding, the following events occur "expression of oligo-adenylate synthetase [2-5 (A) synthetase, activation of RNAase, degradation of viral RNA. In addition, activation of dsRNA-dependent Protein Kinase (PKR), Phosphorylates eIF-2, protein synthesis is inhibited and viral replication is inhibited. IFN signaling activates NK cells and starts eliminating virally infected cells.

Antibody Protection Against Viruses

Antibodies bind to viral surface antigens protect against re-infection and huge amounts of secretory IgA in lumen block viral attachment. Viral entry into cells is mediated through receptors.

Vaccines

The term was coined by Pasteur to honor Jenner's work. Vaccines are cost-effective uses of our immune system, dramatic reduction of Diptheria, Measles, Mumps, Pertussis, Polio and Tetanus. No more naturally acquired cases of smallpox.

For other diseases: TB, malaria and HIV increase the safety of the present vaccine, lower cost, and dissemination. Vaccine development is long and laden with: side effects, exacerbation of disease state and acquisition of disease state.

Immunization

Passive immunity & short-term protection, **Transient** protection (remedy for current problem), involves the transfer of preformed Ig: between ♀

and fetus (trans-placental) & colostrum or by injection. It is given to those exposed to botulism, tetanus, Diptheria, hepatitis, rabies, measles, snake/insect bites. It provides immediate protection to healthcare/travelers. It has passive immunity does not activate and produces no memory.

Risks of Passive Immunity

If the antibody is produced in another spp, the human recipient can produce immune response. In some → IgE production vs isotypic Ab systemic mast cell anaphylaxis. In others IgM or IgG vs isotype complement activation Type III reaction. If human gammaglobulin results can be less severe.

Active Immunization and Long-Term Protection

It promotes protective immunity and memory is achieved by: Natural infection and artificial introduction of whole cells/antigens". The immune system plays an active role to stimulate Ag-reactive T/B cells, Immunizations have played a big role in the decrease of disease in children.

Childhood Vaccines

There are 7 major vaccines such as:

HepB

DTaP

IPV

MMR

Hib

Var

PCV

Types

It is usually requiring repeated boosters. It has predominantly humoral immune response risks of containing active pathogens.

Recombinant Vector Vaccines

Genes are encoding significant Ags from pathogens may be transferred to attenuated viruses/bacteria, include: vaccinia, polio, adenoviruses, salmonella, BCG strain of *M. bovis*. Other vectors may prove to be safer.

Monoclonal Antibodies

Monoclonal antibodies are monospecific antibodies that are "identical because they are produced by one type of immune cell (a single parent cell). Antibodies are produced from a single clone of B cells. It has produced by fusing a B cell secreting the desired antibody with a myeloma cell capable of growing indefinitely in tissue culture. Monoclonal antibodies all have identical antigen-binding sites. They all bind to the same epitope with the same affinity. They are all of the same antibody class (isotype). Polyclonal antibodies are a mixture of antibodies with different antigen binding sites that may bind to different epitopes or antigens with varying affinities. They may be of different antibody classes. The serum obtained from an immunized animal is referred to as a polyclonal antiserum.

Hybridoma

Hybridomas are cells that have been engineered to produce a desired antibody in large amounts. To produce monoclonal antibodies. B-cells are removed from the spleen of an animal that has been challenged with the relevant antigen. These B-cells are then fused with myeloma tumor cells that can grow indefinitely in culture. This fusion is performed by making the cell membranes more permeable. The fused hybrid cells (hybridomas), being cancer cells, will multiply **rapidly** and **indefinitely** and will produce large amounts of the desired antibodies.

Monoclonal Antibody Production

Immunize animal, Isolate spleen cells (containing antibody-producing B cells), Fuse spleen cells with myeloma cells" (eg using PEG - polyethylene glycol). It allows unfused B cells to die. Add **aminopterin** to culture to kill unfused myeloma cells. Clone remaining cells (place 1 cell/well and allow each cell to grow into a clone of cells). The supernatant of each clone for the presence of the desired antibody. It grows the chosen clone of cells in tissue culture indefinitely. We harvest the antibody from the culture supernatant.

Hybridoma Selection

Myeloma cells lack certain enzymes so that they can not use hypoxanthine, aminopterin, and thymidine (**HAT** medium) as a source for nucleic acid biosynthesis and will die in culture. Only B cells that have fused with the engineered myeloma cells will survive in culture when grownn in HAT medium.

Methods of Purification

- Filteration for larger particles
- Ultrafiltration esp. for low concentration samples
- Chromatography:

Ion exchange chromatography (either cation or anion) can be used (Most impurities are usually anions). Size exclusion chromatography: recombinant monoclonal antibodies. The term recombinant and especially recombinant DNA refers to an artificial method of producing DNA (synthetic DNA and proteins). Recombinant DNA is produced through the addition of relevant DNA into an existing organismal genome, such as the plasmid of bacteria. Recombinant antibody engineering involves the use of viruses or yeast to create antibodies, rather than mice. A segment of DNA is recombined (recombinant) with the original DNA of the bacteria, these genetically modified bacteria rapidly reproduce and by that reproduce the DNA segment that was recombined with the original genome.

Uses of Monoclonal ABS

The measuring protein and drug levels in serum. Identifying infectious agents and clusters of differentiation (CDs) for classification and follow-up therapy of leukemias and lymphomas identifying tumor metastasis, identifying and quantifying hormones and immunoaffinity purification.

Chapter 5

INDUSTRIAL MICROBIOLOGY AND FERMENTATION

38. HISTOLOGICAL LANDMARKS, SCREENING OF MICROORGANISMS FOR NEW PRODUCTS

Introduction

Uses microorganisms, typically grown on a large scale, to produce products or carry out chemical transformation. It is originated with alcoholic fermentation processes. Later on, processes such as the production of pharmaceuticals, food additives, enzymes, and chemicals were developed. Major organisms used are fungi and *Streptomyces*. The methods are used to select high-yielding microbial variants.

Properties of a Useful Industrial Microbe

It produces spores or can be easily inoculated. It grows rapidly on a large scale in the inexpensive medium. It produces desired product quickly. It should not be pathogenic. It has amenable to genetic manipulation.

Microbial Products of Industrial Interest

- Microbial cells
- Enzymes

- Antibiotics, steroids, alkaloids
- Food additives

Commodity Chemicals

- Inexpensive chemicals produced in bulk.
- Include ethanol, citric acid, and many others.

Production and Scale

Primary metabolite: produced during exponential growth, eg. alcohol

Secondary metabolite: produced during stationary phase

Secondary Metabolites

It is not essential for growth; formation depends on growth conditions. It is produced as a group of related compounds. It is often significantly overproduced. It is often produced by spore-forming microbes during sporulation. The secondary metabolites are often large organic molecules that require a large number of specific enzymatic steps for production. The synthesis of tetracycline requires at least 72 separate enzymatic steps. Starting materials arise from major biosynthetic pathways. Fermentor is where the microbiology process takes place. Any large-scale reaction is referred to as fermentation.

Most Are Aerobic Processes

Fermentors vary in size from 5 to 500,000 liters, Aerobic and anaerobic fermenters, Large-scale fermentors are almost always stainless steel, Impellers and spargers supply oxygen.

Antibiotics: Isolation, Yield, And Purification

Antibiotics

The compounds that kill or inhibit the growth of other microbes, typically secondary metabolites. Most antibiotics in clinical use are produced by filamentous fungi or actinomycetes, still discovered by laboratory screening. Microbes are obtained from nature in pure culture. The assayed for products that inhibit the growth of test bacteria.

Cross-streak method

It is used to test new microbial isolates for antibiotic production, most isolates produce known antibiotics, most antibiotics fail toxicity and therapeutic tests in animals, time and cost of developing a new antibiotic is approximately 15 years and $1 billion. It involves clinical trials and U.S. FDA approval. Antibiotic purification and extraction often involve elaborate methods.

Industrial Production of Penicillins and Tetracyclines

Penicillins are β-lactam antibiotics. Natural and biosynthetic penicillins, semisynthetic penicillins. It has a broad spectrum of activity; penicillin production is typical of a secondary metabolite. The production only begins after near-exhaustion of carbon source. High levels of glucose then repress penicillin production.

Biosynthesis of Tetracycline Has A Large Number of Enzymatic Steps

- It has more than 72 intermediates.
- It has more than 300 genes involved.
- It has complex biosynthetic regulation.

Vitamins and Amino Acids

The production of vitamins is second only to antibiotics in terms of total pharmaceutical sales. Vitamin B_{12} is produced exclusively by microorganisms. The deficiency results in *pernicious anemia*. Cobalt is present in B_{12}. Riboflavin can also be produced by microbes.

Amino acids

They are used as feed additives in the food industry. It is used as a nutritional supplement in the nutraceutical industry. It is used as starting materials in the chemical industry examples. include, glutamic acid (msg), aspartic acid and phenylalanine (aspartame [nutrasweet] and lysine.

Enzymes as industrial products

Exoenzymes: enzymes that are excreted into the medium instead of being held within the cell; they are extracellular. It can digest insoluble polymers such as cellulose, protein, and starch. Enzymes are useful as industrial catalysts. It produces only one stereoisomer and high substrate specificity and alcoholic beverages and biofuels.

Wine

Brewing and distilling and biofuels. Most wine is made from grapes. Wine fermentation occurs in fermentors ranging in size from 200 to 200,000 liters. Fermentors are made of oak, cement, glass-lined steel, or stone (Figure 15.12b, c, and d). White wine is made from white grapes or red grapes that have had their skin removed. Red wine is aged for months or years. White wine is often sold without aging.

Brewing and distilling

Brewing is the term used to describe the manufacture of alcoholic beverages from malted grains. Yeast is used to produce beer. Two main types of brewery yeast strains. Top fermenting ales. Bottom fermenting lagers. Distilled alcoholic beverages are made by heating previously fermented liquid to a temperature that volatilizes most of the alcohol. Whiskey, rum, brandy, vodka, gin >50,000,000,000 liters of ethanol are produced yearly for industrial purposes. It is used as an industrial solvent and gasoline supplement.

Biofuels

Ethanol biofuels: ethanol is a major industrial commodity chemical. It has over 60 billion liters of alcohol are produced yearly from the fermentation of feedstocks.

Petroleum Biofuels: production of butanol: synthesis of petroleum from green algae.

Products from Genetically Engineered Microorganisms

- Expressing mammalian genes in bacteria.
- The production of genetically engineered somatotropin.
- Other mammalian proteins and products
- Genetically engineered vaccines
- Mining genomes
- Engineering metabolic pathways

Expressing Mammalian Genes in Bacteria

Biotechnology: use of living organisms for industrial or commercial applications, genetically modified organism (GMO): An organism whose genome has been altered, genetic engineering allows expression of eukaryotic genes in prokaryotes (e.g., insulin). This is achieved by cloning the gene via mRNA and finding the gene via the protein.

39. IMPROVEMENTS OF STRAIN, FERMENTATION PROCESS DEVELOPMENT, SHAKE FLASK, SCALE UP OF PROCESS AND BIOREACTORS

Fermentation is a metabolic process that "produces chemical changes in organic substrates through the action of enzymes. In biochemistry, it is defined as the extraction of energy from carbohydrates in the absence of oxygen. Fermentation in food processing is the process of converting carbohydrates to alcohol or organic acids using microorganism's yeasts or bacteria under anaerobic conditions. Fermentation usually implies the desirable action of microorganisms. In microorganisms, fermentation is the primary means of producing adenosine triphosphate (ATP) by the degradation of organic nutrients anaerobically. The science of fermentation is known as zymology. "Since ancient times, humans have exploited the fermentation process. The earliest archaeological evidence of fermentation is 13,000-year-old residues of a beer found in a cave near Haifa in Israel. Another early alcoholic drink made from fruit, rice, and honey, dates from 7000-6600 BC, in the Neolithic Chinese village of Jiahu and wine making dates from 6000 BC in Georgia. Seven-thousand-year-old jars containing the remains of wine, now on display at the University of Pennsylvania were excavated in the Zagros Mountains in Iran. There is strong evidence that people were fermenting alcoholic drinks in Babylon 3000 BC, ancient Egypt 3150 BC, pre-Hispanic Mexico 2000 BC and Sudan 1500 BC.

Traditional fermentation technology is more than 3000-year-old in India as mentioned in ancient literature. A sweet substance known as Soma juice prepared by the Vedic Aryans is supposed to be the first product of fermentation in India. The Rigveda (1500 BC) shows that fermentation technology took its first step in connection with the preparation of Soma juice in India. There is also another drink known as Sura (wine/beer) prepared by fermentation. So, it is believed that acetic fermentation was known to India since the early times. Curd is another very popular fermentation product described in the Rigveda. It still remains a popular food and the technology of curdling milk is also found in a number of texts associated with Yajurveda. In the beginning, fermentation was mainly associated with the preparation of spiritual drinks, but later on, it was used for other purposes also. Finally, the term fermentation is derived from the Latin word Fermentum that stands for boiling. The French chemist Louis Pasteur founded zymology (study of fermentation) when in 1856 he connected yeast to the fermentation process. Food fermentation is the conversion of sugars and other carbohydrates into alcohol, organic acids and carbon dioxide. All three products have shown their importance in human use. The production of alcohol is made use of when fruit juices are converted to wine, when grains are made into beer and when foods rich in starch, such as potatoes, are fermented and then distilled to make spirits such as gin and vodka. So, Fermentation is one of the oldest technologies used for food preservation and it is a process of digesting certain substances that leads to the chemical conversion of organic substances into simpler compounds.

Over the centuries, it has evolved, been refined and diversified. Today a variety of fermented foods is produced both in industrialized and developing countries. A wide range of raw materials is used as substrates in fermentation and resulted in end products derived from fermentation are major constituents of the human diet all over the world. Many benefits are attributed to fermentation. It preserves and enriches food, improves digestibility, and enhances the taste and flavor of foods. Fermentation systems may be liquid, also known as submerged or solid state, also known as surface. Most fermentors used in industry are of the submerged type, because the submerged fermentor saves space and is more susceptible to engineering control and design. Much work still needs to be done to identify the best fermentation technique for each bioactive compound. Fermentation is the chemical transformation of organic substances into simplex compounds by the action of enzymes, complex organic catalysts which are produced by microorganism " such as yeast, molds or bacteria". In other words, fermentation is the technique of biological conversion of complex substrates into simple compounds by various microorganism.

Fermentation, Classified On The Basis Of Substrate

1. Solid state fermentation (SSF)
2. Submerged fermentation (SMF)

The development of these fermentation techniques leads to industrial-level production of bioactive compounds such as antibiotics, pigments, antioxidants, antitumor agent, bio-surfactants, bioactive peptides , etc. The metabolism exhibited by microorganisms is different in SSF and SMF.

Solid-state fermentation (SSF).

Solid-state fermentation (SSF)

Solid-state fermentation (SSF) has been defined as the fermentation process occurring in the absence of free water. Solid-state fermentation is a method used for the production of enzymes, which involves the cultivation of microorganisms on a solid substrate, such as grains, rice and wheat bran. SSF employs natural raw materials as carbon source such as cassava, barley, wheat bran, sugarcane bagasse, various oil cakes like palm kernel cake, soybean cake, groundnut oil cake, fruit pulps (e.g. apple pomace), sawdust, seeds (eg. tamarind and jack fruit), coffee husk and coffee pulp, tea waste etc.

Important Features

Solid state fermentation has been defined as the fermentation process occurring in the absence or near absence of free water utilizing the solid substrate. It is a biomolecule manufacturing process used in the food, pharmaceutical, cosmetic, fuel and textile industries. These biomolecules are mostly metabolites generated by microorganisms grown on a solid support selected for this purpose. This technology for the culture of microorganisms is an alternative to liquid or submerged fermentation, used predominantly for industrial purposes

- Simple and cost-effective
- Less effluent release, reduce pollution
- Aeration is easy
- Resembles the natural habitat of some fungi and bacteria
- SSF utilizes solid substrate, thus nutrient rich waste materials can be easily recycled as substrate
- Substrates are used very slowly and steadily so the same substrate can be used for a longer fermentation period
- SSF is best suited for fermentation techniques involving fungi and microorganisms that require less moisture content

Factors Involved in SSF Process

- Selection of micro-organisms
- Substrate
- Process optimization

Product Isolation & Purification

This is one of the key factors for improved yields of the product Bacteria, yeast and filamentous fungi can be used. Filamentous fungi have shown better results growing in the solid substrate fermentation. Substrate also plays important role in determining the growth of micro-organisms, thereby increasing the product yield. A substrate is chosen such a way that it should provide physical support as well as nutrients to the growing culture. Substrate is of two types: one is specific substrate, which requires suitable value-addition and or disposal. The second is for producing a specific product from a suitable substrate.

Process optimization: It is the optimization of physicochemical and biochemical parameters which includes size, initial moisture, pH and pre-treatment of the substrate, Relative humidity, temperature of incubation, agitation and aeration and age and size of the inoculum. Nutrient supplementation such as N, P and trace elements and supplementation of additional carbon source and inducers. Extraction of product and its purification.

Applications: Solid state fermentation is being employed in various fields ranging from pharmacology to bioremediation, covering various aspects of biodiversity conservation. These are the applications of solid state fermentation. The production of industrial enzymes as almost all the known microbial enzymes can be produced under SSF systems. Enzymes of industrial importance, like proteases, cellulases, ligninases, xylanases, pectinases, amylases, phenolic acid esterases, microbial rennets, oligosaccharide oxidases etc. using SSF systems. The production of Bio-pesticides. The famous *Bacillus thurengenesis* (Bt)'s protein can be produced on a large scale in order to address the issues of pest attacks-yield damage.

Submerged Fermentation

Submerged fermentation / liquid fermentation. It is a technique of cultivation of microorganisms in liquid broth which breaks down the nutrient to release the desired bio-active compound into solution. In this method, the selected microorganism is grown in closed vessels containing a broth rich in nutrients and a high concentration of oxygen. In SMF, substrates are utilized quite rapidly hence need to be constantly replaced or supplemented with nutrients. Bacteria that require high moisture content or high water activity are best suited for submerged fermentation. Applications of Submerged fermentation (SMF)/Liquid Fermentation (LF), SMF utilizes free-flowing liquid substrates, such as molasses and broths. The bioactive compounds are secreted into the fermentation broth. The substrates are utilized quite rapidly that's why substrate needs to be constantly replaced/supplemented with nutrients. This fermentation technique is best suited for microorganisms such as bacteria that require high moisture. An additional advantage of this technique is that the purification of products is easier. SMF is primarily used in the extraction of secondary metabolites that need to be used in liquid form. More than 75% of the industrial enzymes are produced using SMF" one of the major reasons is that SMF supports the utilization of genetically modified organisms to a greater extent than SSF. In SMF, the accumulation of a variety of intermediate metabolites results in lowered enzyme activity and production efficiency. Based on research, certain bioactive compounds have been found to be produced in higher quantities in SSF, whereas other compounds have been extracted using SMF.

40. METABOLITE ANTIBIOTICS ENZYMES, BIOPESTICIDES, BIOREACTORS, ETHANOL AND PRODUCTION

Antibiotic

Natural substances are produced by various species of microorganism's bacteria, fungi, actinomycetes suppress growth and kill other microorganisms.

Antimicrobial Agents

Synthetic analogs and antimicrobial agents are including synthetic as well as naturally obtained drugs that attenuate microorganisms.

Chemotherapeutic Agents

Drugs in this class differ from all others in that they are designed to inhibit/kill the infecting organism and have no/minimal effect on the recipient.

Classification of AMA'S

Microorganisms of medical importance fall into categories are

- Bacteria
- Viruses
- Fungi
- Parasites
- Anti-bacterial
- Anti-viral
- Anti-fungal
- Anti-parasitic agents

Mechanism of Action

- Agents that inhibit the synthesis of bacterial cell walls.
- Penicillins & cephalosporins.
- Cycloserine.
- Vancomycin.
- Bacitracin.
- Azole antifungal agents (clotrimazole, fluconazole, itraconazole).

Agents that act directly on the cell membranes of the "microorganisms such as Polymixin, Polyene antifungal agents (Nystatin, Amphotericin B), Alter cell membrane permeability. Leakage of intracellular compartments. The agents that affect the function of 30S or 50S ribosomal subunits cause reversible inhibition of protein synthesis. Bacteriostatic drugs chloramphenicol, tetracyclines, erythromycin, clindamycin, pristinamycins. Agents that bind to 30S ribosomal subunit & alter protein synthesis, which eventually leads to cell death. Aminoglycosides agent act directly on the cell membrane. Agents that affect bacterial nucleic acid metabolism: Rifamycins which inhibit RNA polymerase and Quinolones which inhibit topoisomerases and anti-metabolites including trimethoprim & sulphonamides.

Antiviral Agents

- Nucleic acid analogs.
- Non-nucleoside reverse transcriptase inhibitors.
- Inhibitors of viral enzymes.

Crop Production and Pest Management

Pesticide product withdrawals, pesticide resistance, zero detectable residues, sustainable food chain: economic, environmental, social and Integrated Pest Management (IPM).

Biopesticides: Mass Produced Biologically Based Agents Used For the Control of Plant Pests

- Living organisms (natural enemies).
- Micro-organisms (Arthropods & nematodes).
- Naturally occurring substances ('biochemicals').
- Plant extracts.
- Semiochemicals (pheromones & allelochemicals).
- Commodity substances.
- Pests = arthropods, plant pathogens & weeds.

Plant Protection Product Registration System

- Two-tier system of registration (legislation under revision in co-decision process).
- Active ingredients at EU level, products at the member state level.

- National authorizations PSD.
- Mutual recognition not working.

Biopesticide PPP Active Substances

- Forty insecticides (27 pheromones, 8 microorganisms, 5 baculovirus).
- Twenty fungicides (all microorganism non-Bt).
- One Nematicide, 6 Repellants (all botanical), 2 others (both botanical).
- It includes substances voted with an entry into force date of 1 May 2009.

Biopesticides and IPM

It is often very specific and compatible with other control agents with "Little or no residue, Inexpensive to develop. The natural enemies used in ecologically-based IPM. Lower potency than synthetic pesticides. But uptake has been low & potential benefits are not yet being realized. Economics (market size, external costs). Efficacy (potency, application, formulation). IPM (integration, best use of biological characteristics). Regulation (system principles, design & operation)". Ecology of insect pathogenic fungi, Genetic structure of natural populations" Ecological factors determining the occurrence of natural populations. Theoretical basis for understanding fate, behavior and environmental impact of biopesticides strains.

Bioreactors

Importance of considering process engineering factors when culturing cells. "Biological factors include the characteristics of the cells, their maximum specific growth rate, yield coefficient, pH range and temperature range. The productivity of a "fermentation is determined by the mode of operation of the fermentation process eg. the advantages of fed-batch and continuous fermentations over batch fermentations. The oxygen demand of an industrial process is generally satisfied by aeration and agitation. Productivity is limited by oxygen availability and therefore it is important to the factors that affect a fermenter's efficiency in supplying O_2. O_2 requirement, quantification of O_2 transfer and factors influencing the transfer of O_2 into solution. Mass transfer, in particular, oxygen transfer is an important factor that determines how a reactor must be designed and operated. The cost was also described as an important consideration. The larger the reactor or the faster the stirrer speed, the greater the costs involved. The rate of oxygen transfers = driving force/resistance eg. resistance to mass transfer from medium to mo`s are complex and may arise from. Diffusion from bulk gas to gas/liquid interface. Solution of gas in liquid interface. Diffusion of dissolved gas to the bulk of the liquid. Transport of dissolved gas to regions of the cell. Diffusion through a stagnant region of liquid surrounding the cell. All these depend on growth/respiration kinetics.

Mass Transfer

One of the most critical factors in the operation of a fermenter is the provision of adequate gas exchange. The majority of fermentation processes are aerobic. Oxygen is the most important gaseous substrate for microbial metabolism, and carbon dioxide is the most important gaseous metabolic product. For oxygen to be transferred from an air bubble to an individual microbe, several independent partial resistances' must be overcome

Solubility of Oxygen

Both components oxygen and glucose must be in solution before they become available to microorganisms. Oxygen is 6000 times less soluble in water than glucose. A saturated oxygen solution contains only10mg dm^{-3} of oxygen. Impossible to add enough oxygen to a microbial culture to satisfy needs for complete respiration". Oxygen must be added during growth at a sufficient rate to satisfy requirements

Oxygen Requirements of Industrial Fermentations

Oxygen demand is dependant on the conversion of Carbon (C) to biomass. Stoichiometry of conversion of oxygen, carbon and nitrogen into biomass has been elucidated". Use these relationships to predict the oxygen demand for fermentation. Darlington (1964) expressed the composition of 100g of dry yeast $C_{3.92} H_{6.5} O_{1.94}$

Factors Affecting Oxygen Demand

Rate of cell Respiration

- Type of respiration (aerobic vs anaerobic)
- Type of substrate (glucose vs methane)

- Type of environment (e.g pH, temp, etc.)
- Surface area/ volume ratio large vs small cells (bacteria v mammalian cells). hyphae, clumps, flocks, pellets, etc. Nature of surface area (type of capsule etc).

Methods of Aeration

A bioreactor is a reactor system used for the culture of microorganisms. They vary in size and complexity from a 10 ml volume in a test tube to computer-controlled fermenters with liquid volumes greater than 100 m^3. They similarly vary in cost from dollars to a few million dollars. There are the following reactors such as :

- Standing cultures
- Shake flasks
- Stirred tank reactors
- Bubble column and airlift reactors
- Fluidized bed reactors

Chapter 6

PRINCIPLE AND WORKING OF BASIC EQUIPMENT'S

41. PH METER

Introduction

A pH meter is an instrument used to measure the pH (acidity and alkalinity) of a solution, of various foods, juices as well as for setting pH of culture media and biochemical activities of microorganisms (Figure 41.1). It is expressed as a number from 0 to 14 categorized into three types: neutral (pH 7), acidic (pH < 7) and alkaline (pH > 7).

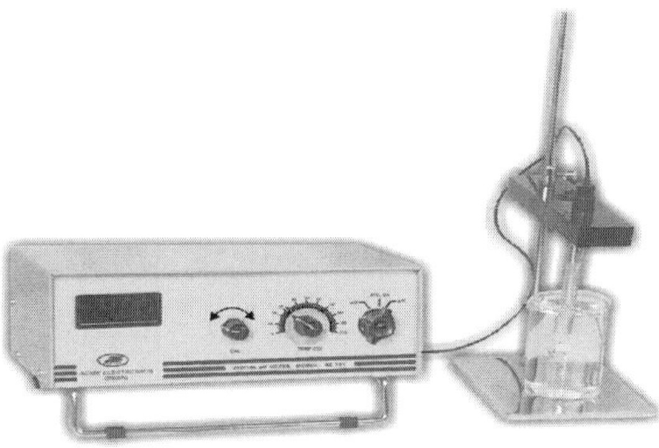

Figure 41.1. pH meter.

Principle

It is expressed in a number from 0 to 14. The number is an expression of the concentration of hydrogen ions (H^+) in the solution- the higher the concentration the more acidic the solution. The value is the negative exponent of the (H^+) concentration. A meter uses an ion-selective electrode (ISE) that ideally responds to only one specific ion, such as the H^+ concentration of the solution. When an ion diffuses from one region of activity to another, there is a free energy charge. The electrode produces a current, proportional to the concentration of H^+ ions, which is converted into a pH reading and displayed on the meter. Based on display type, pH meters are of two types: digital and analog.

A pH meter consists of two separate electrodes: glass and reference (or a single and combination electrodes), H^+ selective membrane and meter with control electronics and display. The round thin glass bulb at the bottom of a pH electrode, which forms the immersion tip in the solution, is the measuring part of the ph meter. The reference electrode is filled with the saturated KCL solutions that complete the measuring circuit and is not sensitive to any of the ions in the solution (Figure 41.2).

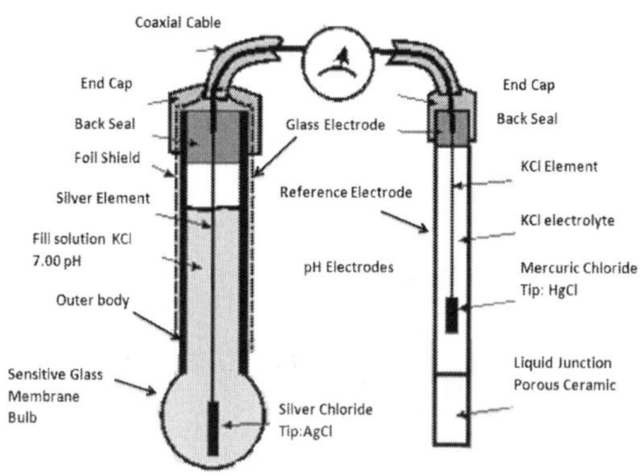

Figure 41.2. Construction of pH electrode.

Procedure

- Calibration or standardization of pH meter
- Select the pH function by pressing the mode function.
- Set the temperature with the temperature knob.
- Remove the electrode from the storage solution rinse with distilled water and blot dry.

- Immerse the electrode in pH 4 buffer and measure its pH.
- Adjust display to read 4 with calibration knob.
- Rinse the electrode and calibrate it with a second buffer of known value (i.e 7 or 10 pH).
- Measure the pH of buffer (pH 4). The pH should read 4.
- The pH meter is now calibrated and is ready to use for measuring the pH of any solution, by the procedure outlined below.
- Remove electrodes from the buffer solution, rinse with distilled water and blot dry.
- Place electrode in the test solution of an unknown pH.
- Turn function sector from the standby to pH
- Read pH of the test solution. Return function selector to stand by.

Precautions

- Calibration should be done at the temperature of the test solution.
- Calibration should be preferably being done every day.
- The electrode should always be kept in distilled water, when not in use.
- Rinsing of electrodes with distilled water is a must before and after moving from one solution to another.

Viva Questions

- What is pH?
- What is the meaning of acidic and alkaline?
- What are the uses of a pH meter?
- What is the principle of a pH meter?
- What are three buffers used for calibrating a pH meter?
- What is the function of KCL?

42. WATER BATH

Introduction

Water bath is an apparatus fitted with a system for the control of temperature and hears the materials by moist heat. However, the use of such naked flame causes serious fire hazards in the laboratories, particularly when any flammable liquid is required heat. In the 1930s Glen and Ruth Morey resolved this problem by invented the heating mantle. The heating mantle is a reliable and non-flammable device used for heating with electric resistance wires woven into a fiberglass cloth sheath. Water baths are mainly of two types: constant temperature water bath and shaking water bath. Water baths are used for heating the solution and those materials that cannot withstand dry heat (Figure 42.1).

These are used for meting of media, testing enzymatic activities, Widal test, Wassermann test for the cultures that require aeration as well as for the extraction of various plant extracts to be tested for pharmaceutical activities. Water baths have multiple usages in laboratories. The medium is used to have the ability to retain heat for a long time. Some of their usages includes (warming various chemicals and reagents, thawing of freeze samples; determination of coliforms, bacteriological examinations, cell cultivation, microbiological assays; corrosion tests; clinical laboratories; food and cosmetic industries and wastewater treatment industries.

Figure 42.1. Water bath.

Principle

It works on the principle of heating of water with a heating element fitted with a thermostat for controlling the temperature and materials are heated with moist heat.

Procedure

- A constant temperature water bath is an insulated metallic vessel fitted with an electric heating mechanism and a thermostat for maintaining the water temperature.
- There are racks for holding test tubes and flasks.
- Water is filled in the container up to the top of the racks and heated at the desired temperature with the thermostat (Figure 42.2).
- A glass beaker with water on a tripod and asbestos with a source of heat beneath (e.g Bunsen burner) can be used as a simple water bath in the laboratory.
- The shaking water bath with a shaker is a thermostatically controlled apparatus. It is used for the cultivation of microorganisms in broth medium.
- It is better than an incubator because it provides a rapid and uniform transfer of heat to the culture vessel and its agitation provides increased aeration, resulting in acceleration of growth and production of enzymes.

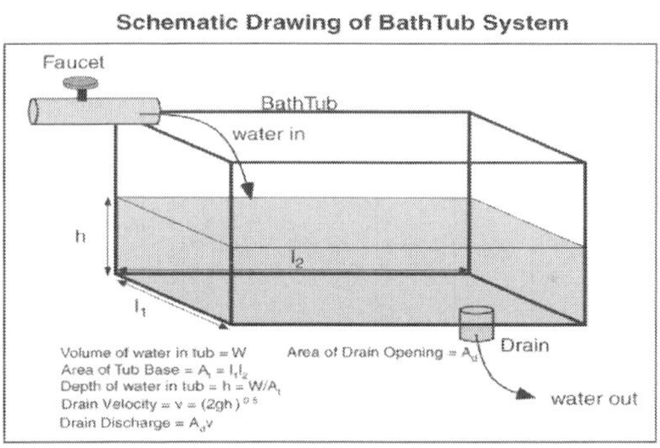

Figure 42.2. Schematic diagram of water bath.

Precaution

- Always maintain the proper water levels in the bath.

Viva Questions

- What is a shaking water bath?
- What are its uses?

- What are the uses of a water bath?
- What is the principle of a water bath?
- How many types of water baths are there?

43. AUTOCLAVE

Introduction

The autoclave is a sealed heating device that uses moist heat under pressure essentially a double jacket steam chamber equipped with devices that permit the chamber to be filled with saturated steam and maintained at a designed temperature and pressure for any period of time. It was developed by Charles Chamberland, one of the Pasteurs associative in 1884. It is a common instrument for very effective sterilization. The pressure increases the boiling water can be sterilized. Steam at a pressure of 15 psi (121°C) kills all organisms. It is the increased temperature and not the increased pressure that kills microorganisms in this procedure. It is used for sterilizing culture media, heat-stable liquids, utensils, linens, bandages, hospital and laboratory equipments not damaged by heat and moisture. According to our ancient history a scientist named Moses, who wrote that the purification of chemicals was achieved by using boiling water and fire. In 460-377 BCE, Hippocrates suggests that the sterilization of surgical instruments can be done by successfully pouring boiling water onto them. After some time in 1729, a famous scientist Lazzaro Spallanzani proved that sterilization of instruments requires only 30 minutes by heating them in a sealed glass (Figure 43.1).

Principle

It is based on the principle where moist heat under pressure at elevated temperature above 100°C effects sterilization. The higher pressure in the autoclave, than higher the temperature of autoclave. Saturated steam heats an object above 2500 times more efficiently than does the hot air at the same temperature. Moist heat kills microbes by coagulating proteins/enzymes (denaturation), which is caused by breakage of the hydrogen bonds. This coagulation process is familiar to any one who has watched an egg white frying.

A simple autoclave is a double- walled cylindrical metallic vessel, made of thick stainless steel or copper, one end of which is open to receive the material (Figure 43.2). The lid of the autoclave is provided with the pressure gauge, steam cock (exhaust valve) controls for pressure adjustment, and safety valve. The basket provides holes all around, for free circulation of the steam is used for keeping the materials. Metallic separations, having holes, are used if more materials are to be sterilized. There are horizontal autoclaves in which pressure, temperature and steam supply are automatically controlled. The steam is heated in the jacket of an autoclave, enters the sterilization chamber through an opening at the upper rear and is expelled through a vent at the bottom front (Figure 43.3).

Figure 43.1. Autoclave.

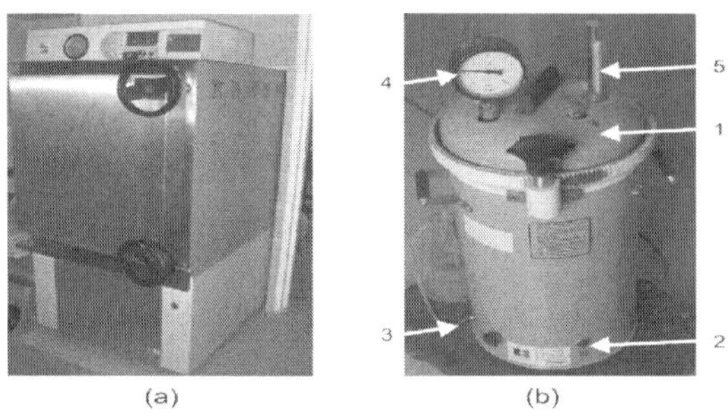

Figure 43.2. Sterilisation by heat the autoclave. (a) Bigger, Automatic Autoclave Operated by External Steam Afflux. (b) Smaller, Manual Autoclave: 1. Lid 2. Power Switch 3. Bleeder Valve 4. Pressure gauge 5. Thermometer.

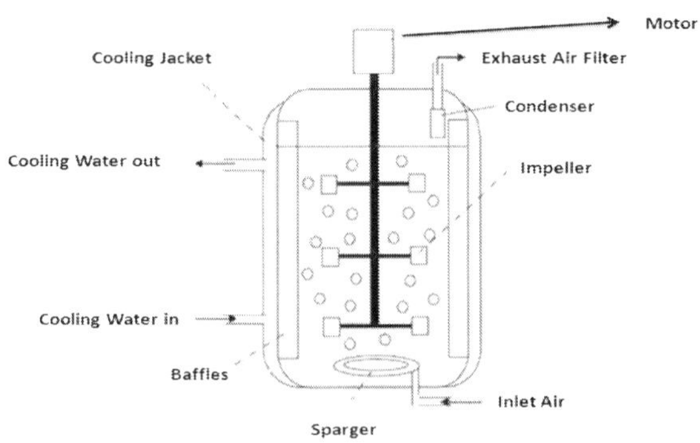

Figure 43.3. A line image of autoclave.

Procedure

- Place the items to be sterilized in the chamber.
- Place the pressurized lid, close and lock the autoclave door by tightening the screw fastener diagonally.
- Switch on the power supply.
- When all the air has been ejected, the temperature reaches 100°C close the operating valve.
- Check the jacket pressure.
- Note the time from the moment temperature reaches 121°C/15lbs pressure.
- Sterilize for 25-20 minutes.
- Switch of the electric supply.
- Allow the autoclave to cool till the pressure falls to the zero mark.
- Unlock and open the autoclave lid.
- Remove autoclave items from the chamber.

Precautions

- The level of water must be checked before operating.
- Articles should be placed loosely so the steam can easily penetrate them.
- Air should be expelled.
- Wrapping objects in aluminum foil is not recommended for dry materials.
- Never open the lid unless the pressure of the chamber comes down.
- Wear insulated gloves when handling hot autoclaved items.
- Cotton plugs should be tightly fitted.

Viva Questions

- What is an autoclave?
- What is the principle of the autoclave?
- What items are not being autoclaved and why?
- What indicator organism is used to test the efficacy of an autoclave?

- What organisms are not killed in an autoclave at 121°C?
- What size of autoclave do we need in a laboratory?
- What kind of loads will be running in an autoclave?
- What is the steam source for the autoclave?
- What utilities are required to prepare for sterilization?
- How many years have the autoclaves been manufactured in the market?

44. BUNSEN BURNER

Introduction

The Bunsen burner name discovered by R.W. Bunsen, is a type of gas burner with which a very hot, practically non-luminous flame, the temperature of which reaches 1870°C at its hottest point, is obtained by allowing the air to enter at the base and mixes with the gas. Direct flaming is one of the simplest methods of dry heat sterilization. It is used in the microbiology laboratory for sterilizing inoculating loops/needles, mouths of culture tubes by incineration and to dry the inside of pipettes (Figure 44.1). Incineration (sterilization procedure by which contaminants are burned to ashes) in a flame is perhaps the most rigorous of all heat treatments for sterilization.

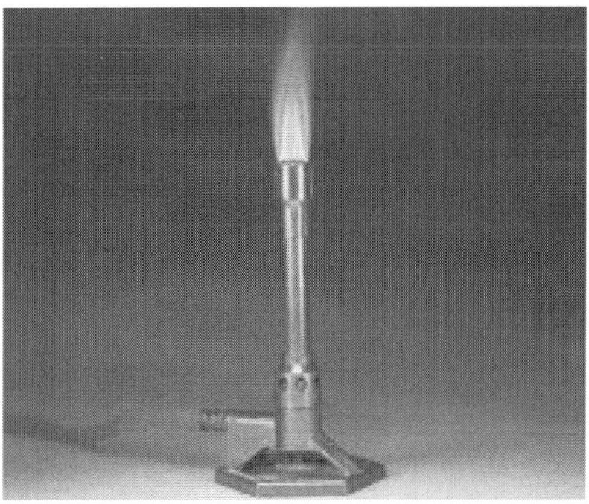

Figure 44.1. Bunsen burner.

Aseptic transfer of a culture from a culture vessel to another is successful only if no contaminating microorganisms are introduced in the process. The transport/transfer of organisms is performed with an inoculating loop or needle. To effectively sterilize, the inoculating loop, prior to picking up the organisms, the end of the wire and the entire loop must be held in the Bunsen burner flame until they glow red hot since the loop is not sterilized by simple moving over a Bunsen burner. The Bunsen burner is providing with the barrel, collar, gastap, air hole, gas jel and base (Figure 44.2)

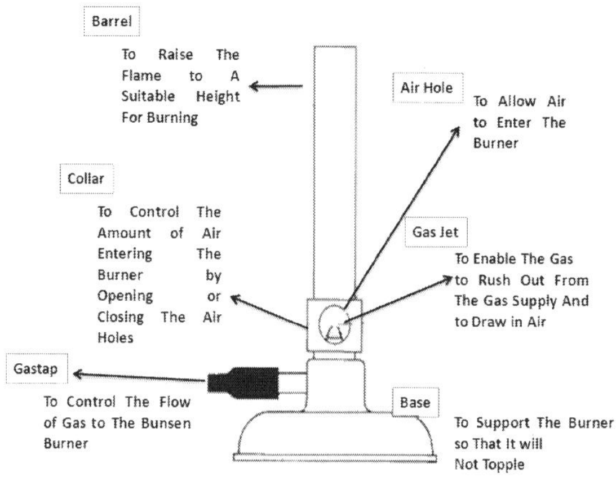

Figure 44.2. Heat the flame: barrel, collar, gastap, air hole, gas gel and base.

Principle

The temperature reaches 1870°C at its hottest point, is obtained by allowing the air to enter at the base and mix with the gas. Direct flaming is one of the simplest methods of dry heat sterilization.

Procedure

- Open the desk outlet value.
- Connect the Bunsen Burner to the gas supply.
- Turn on the gas valve and light the Bunsen burner.
- Adjust the flame, so the blue flame generates an inner blue cone.

Precautions

- Do not allow the formation of floating ashes and aerosols to avoid the spreading of infectious agents.
- Do not close when you hold alcohol in hand.

Viva Questions

- What is the use of the Bunsen burner?
- What is the principle of Bunsen burner?
- What is the use of Bunsen burner in a microbiology laboratory?
- How are the microorganisms killed in the open flame?

45. INCUBATOR

Introduction

The incubator is an apparatus in which media inoculated with microorganisms are cultivated at a constant temperature favorable to the growth of these microorganisms. The very first incubators discovered in Egypt and China that were used in the hatching of fertilized chicken eggs, they consist of fire-heated rooms in which eggs were placed. An incubator similar to a hot air oven consists of an insulted, two walled cabinets, fitted with a heating element at the bottom. It has a thermostat to control the temperature (Figure 45.1). The chamber is provided with perforated shelves adjustable to a convenient height for keeping the items inside.

Figure 45.1. Incubator.

The incubators are provided with double doors, the inner one made of glass, so that the contents and microbial growth in the incubator could be viewed without making the change in the temperature of incubation and growth of the organisms. Both low and high temperatures, then the room temperature, can be maintained in an incubator, depending upon the nature of the microorganism being cultured. Incubators are available in which CO_2/O_2 concentration can be maintained (Fig 45.1). This device helps to maintain optimal humidity, temperature and other gaseous conditions such as oxygen (O_2) and carbon dioxide (CO_2). It is used in microbiology, biotechnology and pharmaceutical industries for culturing microbes at the desired temperature and oxygen requirement.

Principle

It is based on the principle of providing a constant temperature provided through hot-air/dry heat for maintaining the temperature above the room temperature and cool air produced through a cooling system for maintaining the temperature below the room temperature. The temperature of an incubator is kept constant by thermostat control for culturing microorganisms at the desired temperature; the variation of temperature should not be more than one degree Celsius (1°C). While culturing thermophilic microorganisms or incubating cultures for a longer duration, a beaker filled with water (sterilized water) is kept in the incubator to provide moisture or the agar plates are placed in a moist chamber with damp sterile cotton wool at the bottom.

Procedure

- Firstly, start the incubator by providing a power supply.
- Switch "on" the main in the lab.
- Turn on the red color power knob and cooling knob towards 0-1.
- Set temperature of incubator with the help of the driver.
- If the temperature is not shown, check the power supply button.
- The simplest incubators are insulated boxes with an adjustable heater typically going up to 60°C to 65°C (140°F to 150°F), though some can go slightly higher (generally not more than100°C).
- The most commonly used temperature both for bacteria such as the frequently used *E. coli* as well as for mammalian cells is approximately 37°C, as these organisms grow well under such conditions.
- For other organisms used in biological experiments, such as the budding yeast *Saccharomyces cerevisiae*, a growth temperature of 30°C is optimal (for general fungi the optimal temperature is 27°C).

Precautions

- The incubator door should be open only when required.
- It should be taken for not to allow the agar media to dry due to excessive evaporation.
- Simple incubators without a cooling system should be avoided for culturing microbes in the summer season especially in tropical countries.

Viva Questions

- What is the principle of an incubator?
- What is the difference between an incubator and a hot air oven?
- What is the important precaution is taken for culturing thermophilic microbes in an incubator?
- What is the use of an incubator and how many types of incubators

46. HEMOCYTOMETER

Introduction

Hemocytometer is a device used to count cells. It was originally designed for the counting of blood cells. The hemocytometer was invented by Louis-Charles Malassez and consists of a thick glass microscope slide with a rectangular indentation that creates a chamber. This chamber is fixed with a laser-etched grid of perpendicular lines (Figure 46.1). The device is carefully crafted so that the area bounded by the lines is known, and the depth of the chamber is also known. It is therefore possible to count the number of cells or particles in a specific volume of fluid and thereby calculate the concentration of cells in the fluid overall.

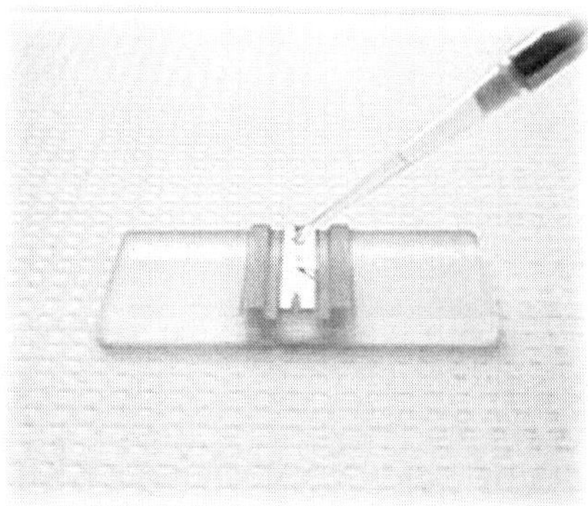

Figure 46.1. Hemacytometer.

Principle

The gridded area of the hemocytometer consists of nine 1 x 1 mm (1 mm^2) squares. These are subdivided in 3 directions; 0.25 x 0.25 mm (0.0625 mm^2), 0.25 x 0.20 mm (0.05 mm^2) and 0.20 x 0.20 mm (0.04 mm^2). The central square is further subdivided into 0.05 x 0.05 mm (0.0025 mm2) squares. The raised edges of the hemocytometer hold the coverslip 0.1 mm off the marked grid, giving each square a defined volume (Figure 46.2).

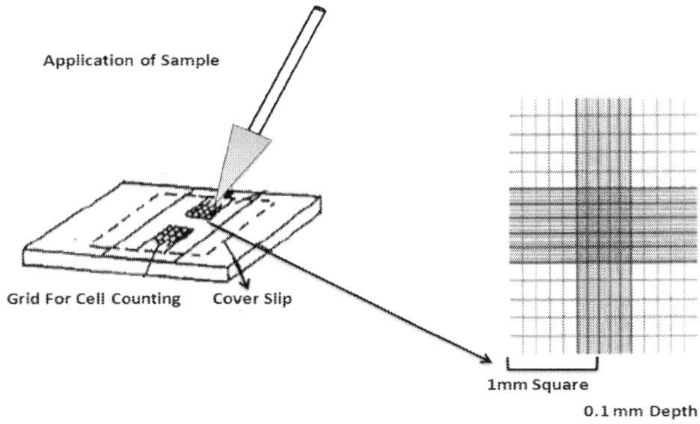

Figure 46.2. Hemocytometer lid.

Dimensions	**Area**	**Volume at 0.1 mm depth**
1 x 1 mm	1 mm^2	100 NL
0.25 x 0.25 mm	0.0625 mm^2	6.25 NL
0.25 x 0.20 mm	0.05 mm^2	5 NL
0.20 x 0.20 mm	0.04 mm^2	4 NL
0.05 x 0.05 mm	0.0025 mm^2	0.25 NL

Procedure

- Take a suspension of cells
- Prepare a 1:1 dilution of the suspension of cells in trypan blue.
- Load the sample in the hemocytometer.
- Cover slip is carefully put on sample side. Overloading of a sample should be avoided.
- Determine the number of cells (total& viable).
- View the cell under a microscope at 100X magnification.
- Calculate the cell no. 4 corner squares and also the average
- Cleaning of the hemocytometer is done after use.

Precautions

- To prepare the counting chamber, the mirror-like polished surface is carefully cleaned with 75% ethanol.

- Use a mechanical suction device to draw up fluids into the pipette.
- Blood and the diluting fluid should be properly mixed.
- Do not allow overflowing of the cell suspension at the time of filling of the cambers.
- The cover slip is also cleaned.
- They must be heavy enough to overcome the surface tension of a drop of liquid.
- A cover slip is placed on the counting surface prior to putting it on the cell suspension.
- Introduce the cell suspension into any of the V-shaped wells with a micropipette.
- The area under the cover slip fills by capillary action.
- Sufficient liquid should be introduced so that the mirrored surface is just covered.
- The loaded counting chamber is placed under the microscope stage and the counting grid is brought into focus at low power.
- Repeat count to check accuracy.

Viva Questions

- What is a hemacytometer?
- What is the number of total quadrants in a chamber?
- What is the surface of a counting chamber?

47. HOT-AIR OVEN

Introduction

The hot air oven consists of an insulated cabinet that is held at a constant temperature by means of an electric heating mechanism and thermostat. A hot air oven is used to sterilize glassware, needles, glass syringes, and metal objects and is the only means of sterilizing oils, waxes and powders. It is a very effective method of sterilization requiring 180°C for 20 minutes, 171°C for one hour, and 160°C for 2 hours or 120°C for 16 hours or longer, depending on the volume (Figure 47.1).

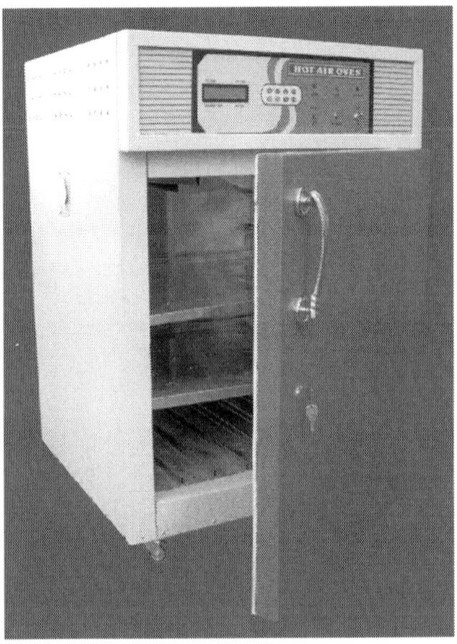

Figure 47.1. Hot Air Oven.

Principle

Sterilization is accomplished by hot-air, a kind of dry heat sterilization. Dry heat penetrates substances more slowly than steam heat. Dry heat is microbicidal and kills the microbes by the oxidation of molecules.

The hot air oven consists of an insulated cabinet, fitted with a thermostat to control the temperature, a fan to keep the hot air circulating at a constant temperature, a thermometer, and perforated shelves for keeping the materials. The heating in an oven is through the coils fitted at the bottom of the cabinet.

Procedure

- Place the items to be sterilized on the shelves.
- Close the door tightly.
- Switch on the power supply.
- Noted time from temperature reaches 120°C or 170°C.
- Sterilize for the desired time period.
- Switch off the power supply.
- Allow it to cool.
- Items will be sterilized and ready to use.

Precautions

- Glassware items should be dried before putting them in the oven.
- Never open the door till the temperature comes down to room temperature.
- To avoid breakage of the glassware items.
- Paper-wrapped items are not to be kept in the oven for sterilization.

Viva Questions

- Explain how dry heat kills microbes?
- What items are sterilized in the oven?
- How does dry heat kill the microbes?

48. VACUUM FILTRATION ASSEMBLY

Introduction

Filtration is the passage of material, liquid or gas through a filter or straining device. It can be used to sterilize substances that are destroyed by heat such as vaccines, some culture media, and enzymes, antibiotic solutions to separate viruses and to collect microorganisms from air and water samples (Figure 48.1).

Figure 48.1. Vacuum filtration assembly.

- **Glass Microfiltration**: Support Systems

Principle

Vacuum filtration is a technique for separating a solid product from a solvent, a liquid reaction mixture, or microbes from a substrate that is poured through a filter paper. The filter traps the solid/microbes and the sterile liquid is drawn through the funnel into the flask below by vacuum. The apparatus used for this purpose is called a filtration assembly. A vacuum is created in the receiving flask; air pressure then forces the liquid through the filter. Membrane filters of different pore sizes ranging from ultrafilters (small molecules), 0.025 μm (smallest viruses, molecules), and 0.45 μm (a few bacteria, viruses, molecules) to 10 μm are used depending upon the particles size.

Requirements

- Filtration assembly
- Filter flask
- Filter papers
- Membrane filter disks
- Disposable dropping bottle
- Vacuum pump
- Sample

Procedure

- Turn on the vacuum.
- Check the vacuum by feeling for suction at the end of the tubing.
- Attach the tubing to the side arm of the filter flask and check the suction at the top of the flask. Put the black rubber ring adapter on the top of the flask and then the Buchner funnel.
- Proper suction to be checked again.
- Place a filter paper/ membrane filter in the Buchner funnel.
- Slowly pour the sample into the funnel.
- Continue to draw air (vacuum) through the sample till it is completed filtered.
- Disconnect the vacuum source.
- Disassemble the membrane filtering unit.
- Decant the filtered sample from the filter flask.

Precautions

- A cotton plug should be used in the vacuum line to ensure sterility
- Always use a sterile filter flask
- Sterile membrane filters should be used.

Viva Questions

- What is the principle of filtration?

- What are the advantages of filtration?
- Where the membrane filtration is used in the microbiology laboratory?
- What substances are to be sterilized through filtration?

49. SPECTROPHOTOMETER

Introduction

Most biochemical experiments involve the measurement of a compound or group of compounds present in a complex mixture. Colorimetry and spectrophotometry are the most widely used methods for determining the concentration of chemical compounds. The basic arrangement of a typical colorimeter is shown in (Figure 49.1).

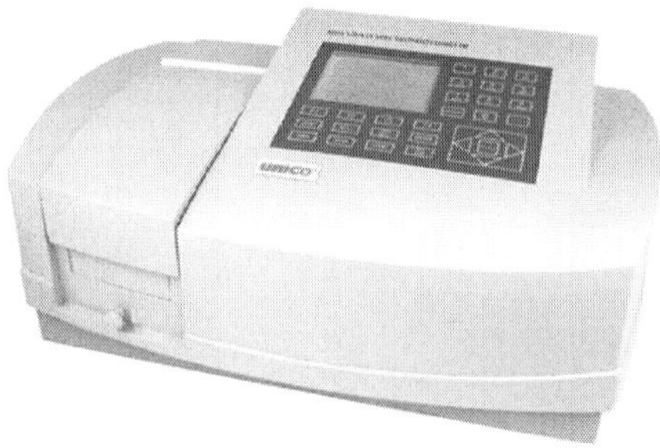

Figure 49.1. Spectrophotometer.

In 1935, Arnold O. Beckman starts working with National Technologies Laboratories, later this company was named Beckman Instruments. In 1940, the first spectrophotometer "Beckman DU spectrophotometer" was sold. It is a device that is commonly used for the measurement of light intensity generated by either reflectance or transmittance of particular substances (solutions, gasses and opaque or transparent). They work in 200 nm- 2500 nm wavelength range of light using different calibrations and controls. Machine calibrations are needed within these ranges of light, by using standards that are of different types depending on the wavelength of the photometric determination.

Principle

A spectrophotometer measures the transmission or absorption of light in liquids as a function of wavelength. The amount of light absorbed is proportional to the intensity of the color and hence to the concentration.

- The amount of light absorbed is proportional to the intensity of the color and hence to the concentration. There are two laws on which the principle of colorimetry is based:

- **The Lambert's Law**-when a ray of monochromatic light passes through an absorbing medium, its intensity decreases exponentially as the length of the absorbing medium increases.
- **The Beer's law**- When a ray of monochromatic light passes through an absorbing medium, its intensity decreases exponentially as the concentrations of the absorbing medium increase. These two laws combined to call Lambert-Beer law.
- When light of a specific wavelength passes through a liquid sample there is a quantitative relationship (Beer's law) between the solute concentration and the intensity of the transmitted light (Figure 9.2), that is,
- $I=I_0 * 10^{-kcl}$
- where I sub 0 is the intensity of transmitted light using the pure solvent, I am the intensity of the transmitted light when the colored compound is added, c is a concentration of the colored compound, l is the distance the light passes through the solution, and k is a constant.

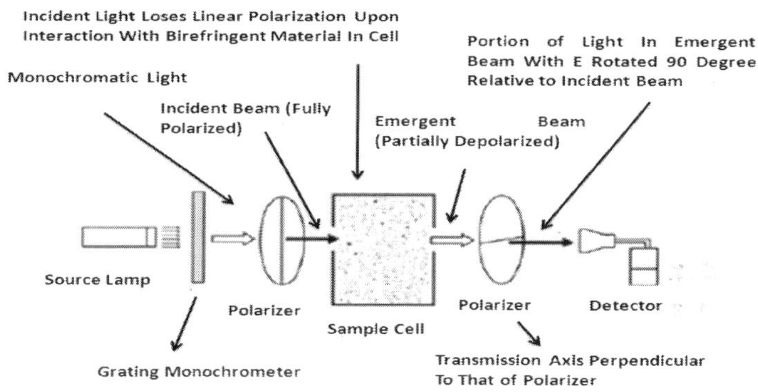

Figure 49.2. Schematic diagram of a modified spectrometer.

A spectrophotometer consists of two parts, *spectron* = for produce light of any selected color (wavelength) and photometer for measuring the intensity of light. In the spectrophotometer, a beam of light is transmitted through a solution or bacterial suspension to a light sensitive detector (Figure 49.3).

Concentration is measured by the extent of absorption of light at the appropriate wavelength registered on the scale of the instrument as the percentage of transmission. Also printed on the instrument scale is a logarithmic expression called the absorbance (also called optical density or OD) which is calculated as: Abs = 2 – log of % transmittance. The absorbance is used to plot bacterial growth. For determining the number of bacteria through turbidity, the bacterial suspension could have more than 10 million cells/ml. Since more than a million cells/ml must be present for the first traces of turbidity to be visible. The higher absorbance limits the greater the concentration of bacterial cells.

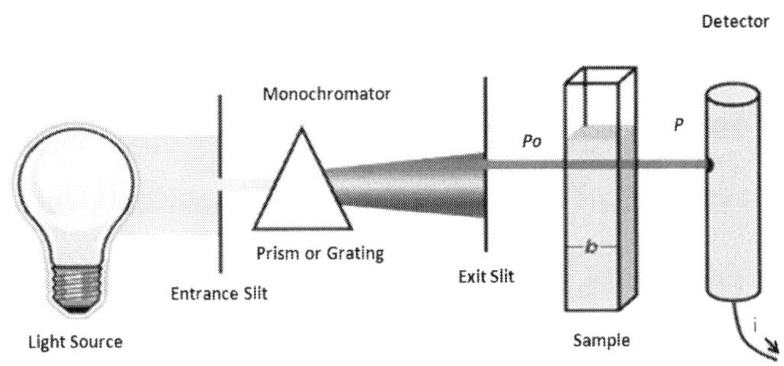

Figure 49.3. UV-VIS spectrometer.

Procedure

- Rinse the cuvette several times with distilled water.
- Switch on the spectrophotometer, 30 minutes before its calibration.
- Rotate the zero control knobs clockwise.
- Set the desired wavelength knob e.g 686 nm for measuring turbidity.
- Adjust the meter needle to zero by rotating the zero control knob.
- Pour 3 ml sterile nutrient broth into the well-rinsed cuvette.
- Open the cover and insert the cuvette into the sample holder.
- Close the cover.
- Keep the index line of the cuvette in line with the index line on the sample holder.
- Rotate the light control knob to adjust the meter to reach 100% transmittance.
- Remove cuvette of nutrient broth and close lid.
- Read, if the needle does not return to zero, readjust accordingly.
- Reinsert nutrient broth cuvette again to see if 100% transmittance still registers.

- Transmittance between 0 and 100% shows the calibrations of the instrument and is ready for turbidity measurements.

Precautions

- Always switch on the instrument 30 minutes before use.
- Don't wipe the cuvette with the towel.
- Finger prints are to be avoided on cuvettes as these will hinder transmittance. Always reset the instrument before using it again for the next sample.
- Handle the cuvettes with great care in order to avoid their breakage.

Viva Questions

- What is the principle of a spectrophotometer?
- What are the uses of a spectrophotometer?
- What is the advantage and disadvantage of estimating bacterial population by measuring turbidity?
- What is the Lambert's and Beer's Law?

50. LAMINAR AIR FLOW BENCH

Introduction

Laminar air flow (LAF) bench or hood is an equipment equipped with a filtered air supply that is free from microorganisms for use in microbiology and tissue culture laboratories, as well as in pharmaceutical industries. It is provided with a special filter system that uses high-efficiency particulate air (HEPA) filters, which is made of boron silicates fibers arranged into a flat sheet.

A laminar air flow cabinet (Figure 50.1) is a closed closet that provides a contamination-free environment. In the hood, air is drawn through HEPA (high-efficiency particulate air) filters and blown in a very smooth, laminar flow towards us. This hood is usually made up of stainless steel with a smooth surface and there are no gaps or joints where spores might collect. These can remove almost all microorganisms larger than 0.3 μm in diameter from the air entering the bench area (Figure 50.1).

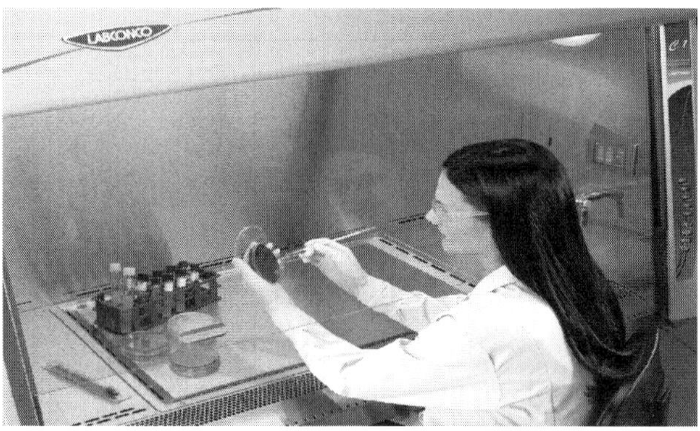

Figure 50.1. Laminar air flow cabinet.

Even though viruses are as small as 30 nm, when air borne they are usually attached to larger particles of dust. It is used for performing the pouring, plating, inoculating, streaking, without any kind of contamination. In the 1960s, the United States' biggest political enemy was the Soviet Union. But in the race to build deadlier nuclear weapons, sometimes the enemy was smaller than one-tenth the size of a human hair. In the immediate post-war era, Sandia focused primarily on weapons development, pushing the limits on miniaturization in both mechanical parts and transistors.

Principle

It works on the principle of air sterilization by filtration. In the laminar flow, a dust free air current flows with uniform and constant velocity along parallel lines in a single direction which help in creating aseptic conditions into the working area. UV light fitted in the chamber kills the air-borne microbes present in the chamber. The principle of a laminar air flow hood is to achieve contamination free environment and protect the work from cross-contamination. Contaminated air enters from inside the front wall to the hood via HEPA filters (filter size 0.3 μm) to the clean bench (Figure 50.2). HEPA filters are considered as the heart of all laminar air flow hoods. This filter removes small particulates generally they are airborne, such as microorganisms. However, it does not provide filtration to vapors or gasses.

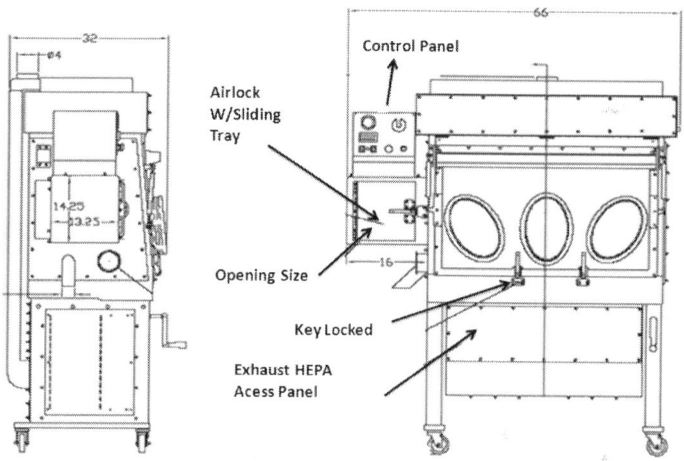

Figure 50.2. Schematic line diagram of laminar air flow (control panel, key locked, exhaust HEPA access panel, opening size and airlock sliding tray).

Procedure

- Switch on the UV light for 30 minutes with the font covering glass still closed, before starting an experiment.
- The covering glass of the LAF bench is opened; the air blower is set at the desired degrees to expel the contaminated air still present in the chamber.
- Now sit properly in front of the chamber.
- Wash hands and also the stage of the chamber with alcohol.
- Place the required materials side by side on the left-hand side of

the stage and now perform experiment within the flame zone of the Bunsen burner/spirit lamp.

- After the experiment is over, stop the motor, wash again the working table with alcohol.

Precautions

- We must wash our hands with detergent.
- We should not talk inside the chamber.
- UV light should never be on while doing an experiment.
- Cleaning of filters at intervals is to be done.
- Avoid using air supply when you are working.

Viva Questions

- What is the principle of a laminar air flow bench?
- What types of filters are used in the LAF bench?
- What is the size of the pores of these filters?
- How has the LAF bench has been named?
- What precautions are needed while using the LAF bench?

51. MICROSCOPY

Introduction

Microorganisms are too much small to be seen with the unaided eye. The study of "microorganisms is greatly enhanced by the use of microscopes. The word microscope, first coined by Giovanni in 1625, derived from the Latin word micro (small) and the greek word scopos (to look at), is an important tool for microbiologists and the technology of making very small things visible to the human eye is called microscopy (Figure 51.1). The simple microscope used by Anton van Leeuwenhoek in the 17th century in 1673 had only one lens and was similar to a magnifying glass that could magnify a microbe 300X.

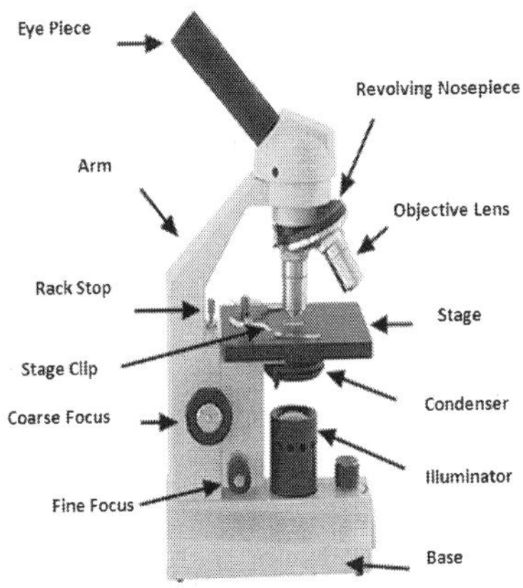

Figure 51.1. Microscope.

The simple microscopes enabled him to be the first person to see bacteria and protozoa. Robert Hooke contemporaries of Leeuwenhoek built compound microscopes that have multiple lenses. The microscopes used today are dividing into two categories: light and electron. Light microscopes are sub-divide into bright-field and fluorescent microscopes, while electron microscope is subdividing into scanning and transmission electron microscopes.

The bright-field microscope or compound light microscope that allows light rays to pass directly through to the eye without being deflected by an intervening opaque plate in the condenser is routinely used in all the microbiology laboratories to study the size, shape, arrangement of microbial cells and sporulating structures.

Dark-field microscopy is a kind of light microscopy that uses a special condenser that contains an opaque disk that causes the light to reflect off the specimen at an angle rather than pass directly through it. This microscopy is used to examine live microorganisms. Fluorescent microscopy uses ultraviolet light instead white light to excite molecules within the specimen or fluorescent dye. Fluorescence microscopy must be intrinsically fluorescent or stained with a dye such as auramine O, acridine orange, fluorescein, DAPI (diamidino-2-phenylinodole).

Bright-Field Light Microscopy

A microscope is a compound optical system, a compound magnifying glass. The real meaning of the functioning of a microscope is a test object positioned between the single and double focus points of the objective. The light coming from the object and passing through the lens creates a magnified, inverted and real image of the subject on the other side of the objective, behind the double focus point.

The eyepiece is at a distance from the objective lens so that the image formed by the objective is generated within the focus of the eyepiece. Thus, looking throughout the eyepieces, one can see a further enlarged, direct but virtual image of this real, inverted and magnified image. The magnification power of a bright-field microscope can be calculated by multiplying the magnification of the objective and of the eyepiece, respectively. The objective lens system consists of multiple lenses. The first member is the front lens facing the object. This determines the magnification and resolution of the microscope. Other elements are responsible for the exclusion of lens errors. The features of an image formed by the objective depend on the optical characteristics of the objective lens.

The quality of a lens represents its property of how sharp the object image is drawn. The image defect, i.e. blurring is caused by lens errors (aberrations). The spherical aberration (spherical divergence) is caused when the lens away from the optical axis increasingly breaks the light rays passing through it, therefore a point-like object in the image will be blurred at the edge. The reason for chromatic aberration is that the focal points of different wavelengths of light do not coincide on the optical axis. The shorter wavelength rays unite closer, while longer wavelength rays unite farther. Thus, a sharp image cannot be obtained using white light, and the image has a colored (rainbow) border. A concave flint glass lens with lead content and high refractive index is fitted to the convex lens to correct for chromatic aberration (the achromatic lens has two-color correction, while the apochromatic lens has three-color correction).

The resolution of the objective lens is the ability of how detailed the image of a subject can be drawn. The resolving power is quantified with the minimum distance between two points that are just distinguishable. The resolution (d) depends on the illumination wavelength of light used (λ), the half-angular aperture of the objective lens (α) and the refractive index of the material between the front lens and the cover slip (n).

$$\mathbf{d = 1.22\ \lambda\ /2\ n \times sin\alpha}$$

where n x sinα = numerical aperture (NA).

The greater the resolution, the smaller the value of d is. This can be achieved by reducing the wavelength of light used, increasing the angular aperture of the objective, or increasing the refractive index of the material between the front lens and the cover slip. Using a light microscope, there is an opportunity only to change the latter. For this purpose, cedar oil (oil immersion) is the most suitable material, because it has almost the same refractive index as that of the glass, so the light passes through a virtually homogeneous medium. The denominator of the above formula, 2 nX sinα is the value of numerical aperture (NA), which may vary between 0.20 and 1.4 (always indicated on the lens). The eyepiece draws a direct image of the test object. The fine structure of an image observed in the microscope depends on the details of the real picture, which in turn is determined by the resolution of the objective. This image, however, is not visible to the naked eye; it can only be visualized in the magnification of the eyepiece.

Microorganisms examine from natural waters by bright-field light microscopy.

Principle

A microscope is a series of finely ground lenses and visible light as a source of illumination is used. The microscope field or area observed is brightly lighted and the objects being studied appear dark because they absorb some of the light. However, microbes absorb light after staining them with a dye greatly increases their light absorbing ability resulting in greater contrast and color differentiation".

Requirements

- Environmental water sample (e.g. from a lake, a stream, a creek or an aquarium)
- Glass slides
- Cover slips
- Glass dropper dispenser alcohol (for sterilization)
- Bunsen burner
- Light microscope

Procedure

- Dry the surface of a glass slide with alcohol over a Bunsen burner and then label the slide
- Place one drop of the sample on the slide with a glass dropper.
- Put the edge of a cover slip on the slide so that it touches the edge of the water drop.
- Slowly lower the cover slip to prevent the formation and trapping of air bubbles.
- Put the sample under the microscope, and locate the focal plane.
- Apply first an objective lens of 16X, and then 40X magnification.
- Observe the shape and movement of microbes, in the case of protists and eukaryotic algae, try to identify the different cell organelles (e.g. chloroplasts, contractile vacuoles)
- Make drawings about the observations.

Fluorescence Microscopy

Specimens that absorb light of color and subsequently emit light of another color (fluoresce) can be visualized by using the fluorescence microscope. The basis of fluorescence microscopy is the principle of the removal of incident illumination by selective absorption, where light absorbed by the specimen and re-emitted at another wavelength is transmitted. The light source must produce light of appropriate wavelength and ineffective wavelengths that are unable to excite the fluorochrome used are removed by an excitation filter. The phenomenon when a sample (e.g. cell organelle of a microbe) shows fluorescence without prior staining is called auto-fluorescence. Otherwise, bacterial cells can be stained with a fluorescent dye. Using a fluorescent dye, e.g. DAPI (4',6-diamidino-2-phenylindole), cells are stained bright blue because the dye forms a complex with the cell's DNA. Fluorescent dyes are therefore widely applied to visualize and enumerate bacteria in different natural habitats or clinical samples.

Microbes examine with DAPI staining (microbes from water sample) DAPI molecules are able to penetrate cell membranes and bind to the double helix of the DNA. Cells can be easily counted, if a known volume of a fixed water sample is filtered through a membrane and, after staining the filter, the surface is investigated by a fluorescence microscope.

Principle

A microscope is a series of finely ground lenses and visible light as a source of illumination are used. The microscope field or area observed is brightly lighted and the objects being studied appear dark because they absorb some of the light. However, microbes do not absorb much light but staining them with a dye greatly increases their light absorbing ability resulting in greater contrast and color differentiation. The effective magnification of a microscope depends upon its resolving power.

- Typical magnification values are:
- Scanning 10X × 4X= 40X magnification
- Low power 10X × 10X = 100X magnification
- High power 10X × 40X = 400X magnification
- Oil immersion 10X ×100X = 1000X magnification
- Total magnification = Magnifying power of the objective lens* magnifying power of the ocular lens.
- The major parts of a microscope and their function.

Requirements

- Disposable gloves laboratory scales paraformaldehyde (PFA) Beaker
- Magnetic stirrer
- Phosphate buffered saline (PBS) NaOH solution
- Pasteur pipette
- Membrane filtration apparatus
- Polycarbonate and cellulose nitrate membrane filters (0.22 or 0.45 µm pore size) 50 ml Falcon tube

- Plastic Petri dishes scalpel
- Pipette with pipette tips DAPI solution (1μg/ml)
- 80% ethanol
- Double distilled water glass slide
- Cover slip
- Vectashield Mounting Medium (H-1000, Vector Laboratories Ltd) immersion oil (non-fluorescent)
- Epifluorescence microscope with a mercury lamp digital camera computer with softwares

Procedure

- For the preparation of the fixative solution, dissolve 1 gm PFA in 50 ml PBS. (PFA causes irritation when inhaled; therefore, the use of a fume hood is recommended.)
- Dissolution can be aided with heating (60°C), permanent stirring and adding some drops of NaOH solution.
- Adjust pH to 7.0.
- Filter the solution through a 0.22 μm pore size membrane filter. (The prepared 2% PFA solution can be stored in the fridge for one week).

Filter the water sample (2-50 ml, depending on the type of sample) using a polycarbonate membrane filter (slowly with occasional stirring).

- To help uniform cell distribution, place a 0.45 μm pore size cellulose nitrate membrane filter between the sieve of the filtration unit and the polycarbonate filter.
- Fill up the Falcon tube with fixative (PFA solution) and immerse the filter in it with sterile forceps (PFA solution must cover the entire membrane filter).
- Incubate the filter overnight at 4°C.
- Fill PBS into an empty Petri dish, then transfer the filter into the PBS solution for 1-2 minutes (liquid must cover the entire membrane filter).
- Transfer the filter to another empty Petri dish and let it dry.
- Cut a 0.5 by 0.5 cm piece from the filter with a scalpel or scissors, and pipette 30 μl DAPI solutions onto its surface. From this step onwards, work in a dark place.

- The filter piece can be marked with a soft pencil.
- After 2 minutes, transfer the filter into 80% ethanol for a few seconds.
- Dip the filter paper into double distilled water for a few seconds.
- Dry the filter.
- Place the filter onto the surface of a glass slide, put a drop of Vectashield Mounting Medium onto the filter and then cover with a cover slip.
- Cover with a paper towel and gently press the cover slip to remove any excess of mounting medium.
- Examine the slide with an epifluorescence microscope using a 100X objective and immersion oil under UV excitation (the absorption maximum of DAPI is at 358 nm, and the emission maximum is at 461 nm).
- Record images from at least 20 different microscopic fields with a digital camera.
- Count the cells on each picture and determine the mean values.
- Determine the cell count for a one ml water sample based on the amount of filtered water and the size of the membrane filter field. Evaluate the variability of cell morphology.

Phase-Contrast Light Microscope

Most of the microorganisms are colorless and also their organelles which often too difficult to see by dark field microscopy and common bright field microscopy. The observations are clear by phase-contrast microscope such as the structure of endospores, unstained bacteria and living organisms. The species are often arranged in pairs or chains, rod shaped and rounded or square ends known as *Bacillus* species and their endospores are mostly oval shape or sometimes cylindrical. *Clostridium* species are repeatedly set in short chains or pairs, mostly with rounded or sometimes pointed ends. The endospores regularly swell up the cell. Consequently, by means of ready slides of *Clostridium* and *Bacillus* the students gain knowledge about a phase-contrast microscope and observing precise bacterial structures such as diverse endospores are present in bacteria. Pond water is generally crowded with protists and bacteria, which observe the internal structure by using a phase-contrast microscope. Apart from this we can also see the internal structure of paramecium (protist) and slowing down the different microorganisms which is present in pond water.

Principle

The colorless and transparent living microorganisms having organelles are generally impossible to see by bright-field microscope due to their absorption, refract and reflect limits. Another reason is sufficient amount of light diffract from the nearby environment for the rest of the microbes. The microbes are visible when they absorb much more light or to refract and reflect from the environment. Thoroughly, we can see microbe organelles by this process. The observations of living and invisible unstained microbes are by using only phase-contrast microscope. The condenser has an annular diaphragm, hollow cone which produces light, glass disk (phase plat) has an objective with a thin film transparent material deposit on it for phase changes produced in the specimen. The phase change of images is observed due to the difference in intensity in light. The positive phase plate is diffracted due to undiffracted light in dark phase-contrast light microscopy or negative phase plate same as above given light relative to the directed light-producing by phase-contrast light microscopy".

Requirements

- Pond wastewater
- Phase-contrast light microscope
- New slides and coverslips
- Pipette
- Microorganisms
- Methyl cellulose "Protoslo, Carolina Biological Supply"
- Tweezers
- Lens paper
- Lens cleaner
- Prepared slides of "*Bacillus* or *Clostridium*"

Procedure

- Create waste pond water.
- Add one drop of methyl cellulose to slow down the swimming of the microorganisms.

- Prepared slides of selected microbes which are given above in the introduction part.
- The slide is placed on the stage of the phase-contrast microscope so that specimen is under the light hole.
- The lenses are set into place and rotate the position (annular diaphragm) of the specimen if it is required.
- It is necessary that the cone of light produced by the annular diaphragm below the condenser be centered.
- Center with the 10X objective and observe the specimen which is under over light.
- Rotate the annular diaphragm into the correct position for the observation with a 40X objective if the image is not seen clearly.
- Do the same with the oil immersion lens.
- In the report, sketch several of the microorganisms that you have observed.
- If you examine pond water (collection of microbes), use the pictorial guides provided by your teacher to help you in identifying microorganisms that are present in pond water.

Precautions

- Make sure the specimen is directly over the light hole in the phase-contrast microscope.
- The phase elements must be properly associated.
- Misalignment is the major consequence that beginning students encounter in phase-contrast microscopy.
- If your microscope is not properly worked ask your lab technician for help.
- Keep the microscope covered when not in use
- I should be kept away from moisture and direct sunlight.
- Never clean lenses with a paper towel or cloth.
- Never use a slide without cover glass while examining an object.

Viva Questions

- What is a phase-contrast microscope?
- What is the principle of this test?

- Why do we use a phase-contrast microscope instead of a bright-field microscope?
- What is the significance of this test?
- What is paramecium?
- What is protista? Why don't we use a bright-field microscope to see them?
- What is a microscope?
- What is the maximum resolution of a microscope?
- What is the principle of a microscope?
- What is the meaning of magnification power?
- What is resolving power?
- Who discovered the microscope?

Chapter 7

EXPERIMENTATION

1. ISOLATION OF FUNGI BY THE WARCUP METHOD

Introduction

Microorganisms are present in our environment. They occur in the air, soil, water, sewage, decomposition matter, living plants, food and our body surfaces. Microbes are encountered by the microbiologist in the microscopic forms of life ranging from viruses, bacteria, mycoplasma, algae, fungi and protozoa to worms. They can be saprophytic, parasitic and capable of living within or outside the cells of higher forms of life or some lower forms of life. The microbiologist hasseparated these microorganisms individually for study. For isolation, several media are used due to their different requirement of nutrients for the growth of microbes. The important precautions to be taken in order to get satisfactory isolated colonies during the performance such as to avoid open windows, random movement of hands, heating system, even talking and contact between the sample and all non-sterile objects and surfaces.

The isolation and identification of soil fungi is an example in which a mycological ecosystem can be studied. It is generally very difficult to isolate and separate the fungi from soil by directly picking them to place them under a microscope for observation. Hence many techniques have been devised for stimulating the growth and easing fungal isolation. This method was developed by Warcup in 1950.

Principle

The Warcup soil plate method is superior to the dilution plate method in that it is easier to use and it is less time-consuming, requires less glassware and get more qualitative information about fungi.

Requirements

- Spirit lamp
- Inoculating loop or needle
- Soil sample
- Sabouuraud agar medium (100 ml)
- Streptopenicillin
- Syringe with needle
- Rose bengal

Procedure

- Collect soil at random in sterilized polythene bags, make a composite sample by mixing the above samples.
- Add 0.005 to 0.15 gm air dried soil each in 5 sterile Petri plates with the help of a sterilized cooled loop or transfer needle.
- Add 15-20 ml of melted, cooled (45°C) medium, supplement with streptopenicillin and rose Bengal, to each soil inoculated Petri plate.
- Dispense soil particles throughout the medium by gentle rotation of Petri dishes.
- Allow the plates to solidify.
- Incubate the plates to solidify.
- Incubate the plates at 25°C in an inverted position for 15 days.
- Observe the plates for the appearance of the fungal colonies after 3 days of incubation.

Precautions

- Transfer needle or loop should be cooled either in sterile medium or in soil.
- Streptopenicillin should not be added to a medium that is too hot, it is always added when prior to pouring the medium into the plates.
- Plates are not placed in an inverted position for incubation.
- Sterile glassware should be used.

Viva Questions

- What are ecosystems?
- What is ecology?
- Differentiate about the fungi and bacteria?
- What is serial dilution?
- Who discovered isolating method in the 1950s?

2. VASCULAR-ARBUSCULAR MYCORRHIZAL (VAM) SPORES ISOLATE FROM SOIL

Introduction

A symbiotic relation is in between hyphae of a soil borne fungus and root of plant which called mycorrhiza. These are the following types such as ecomycorrhiza (growth of the fungus oocurs on the surface of the roots), endomycorrhia (fungi occurs in the roots), ectendomycorrhiza (fungus form a hartig net but also penetrates the roots), vesicular arbuscular mycorrhiza (VAM) the penetrating hyphae of the fungus in the cortical cells may be coiled or branched.

Principle

The vesicular-arbuscular mycorrhizal fungi belonging to the family Endogonaceae are among the more common and widely distributed of the soil-borne fungi. Spores of VAM fungi from soil are isolated by various techniques such as floatation-bubbling method, floatation method, sucrose centrifugation, separation of spores by gelatin column, airstream fractionation and various others method but sieving and flotation methods are the most commonly used to identify VAM fungi.

Floatation Method

Requirements

- Centrifuge tubes
- Centrifuge
- Fine sieve
- Blender
- Sucrose solutions
- Soil samples

Procedure

- Take 20 gm of soil into a blender.
- Blend the blender at high speed for 1 to 2 minutes to free spores attached to the roots, or in sporocarps, or free from clay particles.
- Pass contents of the blender through a finc sicvc and wash the colloidal material thoroughly with a strong stream of water.

- Add 10 ml of 20% sucrose into a clean centrifuge tube, followed by 10 ml of 40 % and then 10 ml of 60% sucrose into the bottom of the tube.
- Add 10-15 ml of blended sieving onto the surface of the 20% sucrose layer.
- Centrifuge for 3 min at 300 rpm.
- Remove debris that collects 20-4-% and 40-60% interfaces.
- Rinse spores on a fine sieve with a strong stream of water to remove sucrose and collect the spores.
- The layer of VAM spores will be visible at 20-40% and 40-60% interfaces that are removed in the last step.

Sieving Method

Requirements

- Bunsen burner
- Soil sample
- 500-ml beaker
- Sieves of 710 μm, 250 μm, 75 μm, 45 μm

Procedure

- Take 200 ml of distilled water into the 500 ml of the beaker.
- Heat the water to 40-50°C for dilution.
- Add 50 gm of soil and mix well until the soil aggregates disperse to leave a uniform suspension.
- Allow the heavier particles to settle down.
- Decant most of the suspension through a 710 μm sieve to remove large organic matter and roots.
- Add 200 ml of water to the suspension.
- Decant the suspension through a 710 μm sieve.
- Decant this through 250 μm, 75 μm, 45 μm sieves.
- Collect the residue on the 45 μm sieve.
- Wash the residue with distilled water and collect the spores.

Precautions

- The aseptic condition should be maintained.

- Wear gloves and a lab coat before the start of the experiment.
- Care should be taken with the centrifuge.
- Sterile distilled water should be used.

Viva Questions

- What is VAM?
- How many types of VAM are there?
- Which method is easy and less time-consuming for the isolation of the VAM fungi?
- What is the significant role of this test?
- What are hyphae?
- Explain the various techniques to be isolated fungi?

3. AZOBACTER SPECIES ISOLATION FROM SOIL

Introduction

It is free-living nitrogen-fixing and aerobic bacteria, which is known as non-symbiotic nitrogen fixer. Non-symbiotic nitrogen fixation is restricted to certain microorganisms mostly bacteria and blue-green algae. The aerobic bacteria are capable of fixing nitrogen belong to entirely different families and have been isolated from different habitats such as soil, fresh and marine waters, animal and others. The common species of Azotobacter are *Azotobacter chroococcum, A. beijerinkii and A. paspali, A. macrocystogenes.*

Principle

It is characterized by the relatively large size of the individual cells as seen in wet preparations under the phase -ontrast microscope. It is often motile with peritrichous or polar flagella and production of thicked walled-microcysts in some species. These are obligate aerobes and capable of fixing atmospheric nitrogen when provided with suitable carbohydrates or energy sources. It can be isolated by the soil dilution method, soil plate method or direct isolation on any of the mediums.

Requirements

- Soil samples
- Bunsen burner
- Marker
- Ashby's medium
- 90- ml sterile water blanks
- Sterile 10 ml pipettes
- Magnet shaker
- **Preparation of Ashby's medium**
  - Mannitol 20 gm
  - K_2HPO_4 0.2 gm
  - $MgSO_4.7H_2O$ 0.2 gm
  - NaCl 0.2 gm
  - K_2SO_4 0.1 gm
  - $CaCO_3$ 5 gm

- Agar 15 gm
- Sterile distilled water 1000 ml

Procedure

- Pour media into beakers and autoclave for sterilization.
- After cooling,pour into the sterile Petri plates.
- Allow the medium to solidify
- Add 10 gm of sieved soil to 90 ml sterile distilled water blank and shake it for 15-20 minutes on a magnetic shaker.
- Prepare serial dilutions 10-^{2}, 10-^{3}, 10-^{4} and 10-^{5}.
- Add 1 ml aliquots of different dilutions and solidify the agar medium in Petri plates.
- Rotate the plates for uniform distribution of spores.
- Incubate the plates at 28°C for 3 days.
- Observe the plates after 3 days of incubation for the appearance of colonies.
- Azotobacter colonies have soft, flat milky and mucoid shape colonies.

Precautions

- Spread the suspension over the medium with the help of a spreader.
- Cooled agar medium should be used for pouring the suspension.

Viva Questions

- What is azotobacter?
- Which media are used for the isolation?
- Explain common species of Azotobacter?
- What is heterotrophic?

4. ALGAE ISOLATION FROM SOIL

Introduction

There are many algae that are found in nature. They occur in the oceans, salt lakes, seas, freshwater, marine water, streams, ponds, damp soils, on rocks, tree barks, and stones and in plants and animals. They are very small in size and smaller than bacteria and fungi. They are either prokryotic or eukaryotic.

Principle

The convenient techniques are used for the isolation of alage; Beneck's broth is commonly used for their isolation.

Requirements

- Soil samples
- Sterile petri plates
- Agar powder
- Half pint bottles
- Laminar air flow instrument
- Platinium or nichrome wire-loop
- Bunsen burner
- Constituents of Beneck's broth

KNO_3 0.2 gm

$FeCl_3$ (1%) 2 drops

$CaCl_2$ 0.1 gm

KH_2PO_4 or K_2HPO_4 0.2 gm

$MgSO_4.7H_2O$ 0.2 gm

This medium is also prepared by the addition of agar (15-20 gm) per liter to the above constituents. The pH should be in between 7.0- 7.5.

Procedure

- Pour the broth in half pint bottles, filled to a depth of 5 cm and plug with cotton-wool.
- Broth is autoclaved for 20 minutes.

- Allow the medium to cool.
- Add 1 gm of soil to each bottle.
- Incubate these at 30-35°C in a growth chamber for 15-20 days.
- After 15 days' incubation observe the bottles for algal growth.
- Algal colonies visualized in the bottles are purified by aseptically transferring a few cells or filamentous of a platinum or nicrome wire loop on agar medium in petri plates to obtain pure colonies.
- After that subculture should be done with identifying of the algal conies.

Precautions

- Phosphate should be autoclaved separately.
- The light intensity should be one-third of the normal sunlight.
- Sterile distilled water should be used.

Viva Questions

- What are algae?
- Where are algae found?
- Which medium should be used for growth?
- Which broth should be used for the isolation of algae?

5. PROTOZOAN ISOLATION FROM SOIL

Introduction

Protozoa are unicellular nophotosynthetic organisms with a eukaryotic cell structure. It is different from fungi by the absence of cell wall. These belong to protista. There are over 40,000 species of protozoa found in oceans, freshwater or damp soil. There are few species which are parasites and are responsible for causing diseases in man such as malaria. The malaria is caused by *Plasmodium vivax*, *Plasmodium falciparum*, *Plasmodium malariae* and *Plasmodium ovale*. They cause genitourinary tract infections (*Trichomonas vaginalis*) and other intestinal infections (*Giardia lambia, Balantidium coli*).

Principle

The protozoa are especially prevalent in tropical and subtropical enviorments. The free-living protozoa are critical links in the food chain, feeding on other microorganisms and subsequently serving as a nutrient-source for higher organisms. Those protozoa are also responsible for the termite's ability to utilize food via digesting of cellulose.

Requirements

- Sterile pipettes
- Soil sample
- Sterile bottles
- Nutrient agar slants of *Aerobacter sp*. 2-5 days old.
- Water agar
- Sterile petri dishes
- Spreader

Procedure

- Pour agar medium into sterile Petri plates.
- Allow the agar to solidify.
- Spread a thick suspension with the help of inoculating loop in the form of circular patch on the agar surface.
- Mark 3 circles in each plate.

- Place a small amount of soil on the center of each circle.
- Moist the soil with a few drops of sterile tap water.
- Put the plates into incubation at room temperature for 2 weeks.
- Add sterile tap water to the soil at frequent intervals to avoid dryness.
- Observe the plates microscopically under low-power objective for protozoans.
- Protozoa will be visible on agar surface that move by phagocytic action with the help of flagella.

Precautions

- Sterile distilled water should be used.
- Aseptic conditions should be maintained.

6. ISOLATION OF MICROORGANISMS FROM RHIZOSPHERE

Introduction

The rhizosphere term was introduced by the German scientist Hiltner in 1904 to denote the region of soil that is under the root. It is defined that the soil immediately surrounding the roots of a plant together with the root surfaces. The region provides certain characteristics conditions for the increased occurrence of microflora in it that is attributed to the rich food materials provided by the added sloughed off portions of root tissues and root exudates. The root exudates contain sugars, amino acids, vitamins and other growth factors that serve the nutrients for microorganisms in the rhizosphere.

Principle

The various techniques have been discovered such as contact slide technique, use of fluorosence microscopy, fluorescent antibody technique, soil dilution and plate count method, soil plate technique and soil box technique to study the rhizosphere microflora. The plate count method and soil dilution are mostly used for the isolation of rhizosphere.

Requirements

- Marker
- Freshly collected roots
- Bunsen burner
- Sterile ploythene bags
- Nutrient agar medium
- Nutrient agar plates
- Czapek-Dox agar medium
- Sterile 100 ml water blanks (250 ml flask) 2
- Sterile 90 ml water blanks (250 ml flasks) 12
- Sterile polythene bags
- Sterile Petri dishes
- Sterile distilled water

- Sterile pipettes
- Screw cap bottle

Procedure

Soil and root samples collected from uncultivated plots in polythene bags and bring to the laboratory.

(A) **Isolation of rhizosphere microflora**

1. The seperation of rhizosphere soil have done from 5-6 roots with the help of a brush in a petri dish.
2. Add 10 gm of soil in 100 ml sterile distilled water blank and shake it for 15 minutes with the help of magnetic shaker.
3. Prepare serial dilutions 10^{-2} to 10^{-6} as given the experiment before.
4. Transfer 1 ml of each dilutions 10^{-4} to 10^{-6} into sterile petri plates (3 replicates) and (6 replicates for 10^{-5}).
5. Pour the Czapek-Dox agar supplemented with strptopenicillin and nutrient agar media for fungi (10^{-4}, 10^{-5}), bacteria and actinomycetes (10^{-5}, 10^{-6}) respectively to various plates.
6. Incubate the CDA plates at 25°C in an inverted position for 7 days and NA plates at 35°C for 24-48 hours.

(B) **Isolation of root surface microflora**

1. Take a root piece from the first dilution in screw cap bottle with 100 ml sterile water and shake it well.
2. Serial washings (10-20) are given to the above roots with sterilized water until clear root surface is exposed.
3. Plate the washed roots on Czapek-Dox agar and nutrient agar plates. Incubate the plates for 7 days at 25°C in an inverted position for 7 days.

(C) Isolation of microorganisms from the non-rhizosphere soil

1. The plates are observed after 2 days of incubation for the appearance of colonies of bacteria and actinomycetes and after the 4-5 days for fungi and continue observing till 2 weeks.
2. Record the data for the number of colonies per plate in various dilutions for bacteria, actinomycetes and fungi in rhizosphere, rhizoplane and non-rhizosphere soils.
3. Identify the various isolates up to generic/specific level and maintain the record for each.
4. Record the surface microflora qualitatively.

Microbial counts in rhizosphere and non-rhizosphere soils per gm of the soil are calculated by applying the formula:

$$\text{Number of microorganisms / gm} = \frac{\text{Number of colonies/ plate x dilution factor}}{\text{Dry wt. of the soil taken}}$$

Rhizosphere effect (R: S) can be calculated by applying the formula:

$$\text{R: S} = \frac{\text{Number of microorganisms in the rhizosphere soil}}{\text{Number of microorganisms in the control soil}}$$

Precautions

- Wear a labcoat and gloves when handling with rhizospheric soil.
- Aseptical experiments should be done.
- Sterile distilled water should be used.

7. ISOLATION OF CYANOBACTERIA (BLUE GREEN ALGAE) FROM PADDY FIELDS

Introduction

Cyanobacteria are a structurally diverse group of gram-negative bacteria characterized by their ability to perform oxygenic photosynthesis and biological nitrogen fixation. The motility occurs by gliding. They do not require growth factors for their growth, excepting marine cyanobacteria which require vitamin B_{12}. They occur in the sea and freshwater and soil. It is also thermophilic and grows abundantly in neutral or alkaline hot springs and can grow at a temperature up to 75°C. The common nitrogen-fixing blue-green algae recorded from rice fields include species of *Anabaena, Aulosira, Cylindrospermum, Clothrix, Scytonema, Tolypothrix, Fischerella, Mastigocladus, Campylonema, Westiella and Nostoc*.

Principle

Blue-green algal cells can be identified after preparing suitable mounts under the low-high power of magnification.

Requirements

- Soil sample
- Bunsen burner
- Inoculating loop
- Conical flasks (100 ml)
- Polythene bags
- Sterile water blanks (90 ml) (5)
- Sterile pipettes
- Sterile water
- Growth chamber
- Constituents of Pringsheim's broth

KNO_3 0.02%

$MgSO_4.7H_2O$ 0.001%

$(NH_4)_2HPO_4$ 0.002%

$CaCl_2.6H_2O$ 0.0005%

$FeCl_3$ 0.00005%

Procedure

- Collect the blue-green algal colonies floating on flooded rice fields in glass bottles or soil sample in polythene bags and bring to the laboratory.
- Pour the broth in conical flasks of 100 ml capacity (40 ml each), plugged with cotton wool and autoclave at 121°C for 2 minutes.
- Prepare serial dilutions of soil (10^{-2} to 10^{-7}) with water blanks.
- Transfer 1 ml of each suspension from the different dilutions into the flasks each containing 20 ml broth.
- Incubate these flasks in the growth chamber for several weeks at 28-32°C.
- Transfer the algal colonies as soon as they are observed either for enrichment or fresh aliquots of liquid media or agar slants.
- Observe the colonies appearing in the broth after time interval on the agar plate (subculturing).

Precautions

- Sterile distilled water should be used.
- Aseptic conditions should be maintained.

8. ISOLATION OF AQUATIC FUNGI

Introduction

The mixture of fungi and spores are found in watercourses, polluted water and sewage. The freshwater has chief fungi which are called as *Chytridiomycetes and Oomycetes*, especially *Chytridiales and Sapropegniales*. A number of them are parasites or freshwater planktons. There are a few freshwater ascomycetes known while other hyphomycetes are quite good in number. They are varying greatly in their requirements for growth and each must adapt to great changes in the environment. These can be extracted from pond or river sewage using selective techniques. The aquatic fungi are complex when there is a mixing of strong parasites of active saprophytes with resting saprophytes suitable substrate is called zymogenous.

Principle

The aquatic fungi from various environments can be isolated using selective techniques. There are two techniques that are commonly used baiting in the field and baiting in the laboratory.

Isolation of fungi by baiting in the field

Requirements

- Sterile distilled water
- Sterile Petri dishes
- Nylon rope
- Baits (Small pears, hard green apples, loquates, plums, tomatoes, grapes, bananas, mangoes, berries, sweet corn kernels, cellophane, oak, twigs, elm, ash, alder, birch, sycamore, human hair, etc).
- Nylon bags

Procedure

- Hang the selected bait and place in nylon bags or in metal cages in the water.
- Withdraw the samples i.e baits at regular intervals from the river or sea and brought to the laboratory.
- Wash the samples in sterile distilled water.

- Transfer the washed baits to sterile Petri plates containing sterile water with fresh baits.
- Incubate in the dark for 2-3 weeks at 5-20°C.

Isolation of fungi by baiting in the laboratory

Requirements

- Water samples in sterile bottles
- Baits (hemp, mustard seeds, grass leaves, house or drosophila flies, termite, wings cellophane, pollen grains of conifers, snakeskin, white human hair).
- Bunsen burner
- Beakers (500 ml)
- Forceps
- Sterile glass distilled water
- Sterile Petri dishes
- Potato dextrose agar medium
- Forceps

Procedure

- Collect water sample, soil, plant and animal debris in sterile bottles and bring to the laboratory.
- Put a few ml of collected water or a small amount of soil or debris in sterile Petri dishes (5-10) which are half-filled with cool, sterilized glass distilled water.
- Add suitable baits to the above Petri dishes.
- Incubate the plates at 5-20°C for two to three weeks in dark.
- Change the water in Petri plates frequently with sterile glass distilled water to avoid a profusion of bacteria and protozoans.
- As soon as the growth of a fungus is seen on bait, transfer the bait to a new Petri plate containing fresh bait and sterile water.
- Transfer the bait with fungus growth on an agar medium and incubate the Petri plates at 20°C.
- Observe the baits with fungus growth under low power magnification.

Precautions

- Change the water frequently to avoid anaerobic conditions, avoiding protozoans and bacteria growth.
- Sterile Petri plates should be used to avoid other contamination.
- Sterile distilled water should be used.
- Incubate the Petri dishes in dark to discourage algal blooms.

9. ISOLATION OF SLIME MOLDS

Introduction

The slime molds are characterized by the presence of a trophic phase that may be an amoeba and aggregation of an amoeba called pseudoplasmodium or plasmodium and belong to the division Myxomycota. They differ from the true fungi in their phagotrophic nutrition and somatic parts devoid of cell walls :consist only of protoplasts bounded by the plasma membranes. They are experimental tools that are useful for geneticists, cytologists, molecular biologists, morphogenetics and biochemists due to cellular slime mold often being one of the simplest systems. The slime mold has ability to undergo a transition from independent cells to a multicellular stage that is known as *Dictyostelium discoideum*.

Principle

The slime mold has not been grown axenically on a chemically defined medium, they only grow on glucose-peptone agar. The amoeba present in a portion of the translucent spot can be separated as a pellet by centrifugation after adding cold water. They are grown in cultures with the association of a bacterium generally with *Enterobacter aerogenes or Escherichia coli* on which they grow and feed.

Requirements

- Soil samples
- Decomposing leaf litter
- Peptone yeast extract dextrose agar medium (30 ml).
- Bunsen burner
- Sterile Petri plates
- Sterile distilled water

Procedure

- Pour the medium into the agar plates.
- Allow the medium to solidify.
- Sprinkle soil particles or finely divided decomposing leaf litter on the surface of the medium.
- Incubate the plates at room temperature for 3-7 days at 20-24°C.

- Examine the plates after 3 days and observe the appearance of slime molds under the microscope.
- Slime molds grow forming patches and wheel like pseudoplasmodia.

Precautions

- Sterile distilled water should be used for making chemicals.
- Aseptic conditions should be maintained.

10. ISOLATION OF MYCOPLASMAS

Introduction

Mycoplasmas were discovered by E. Nocard and E.R. Roux, in 1898. They isolated some microorganisms from the pleural fluids of cattle suffering from bovine pleuropneumonia and were named as pleuropneumonia-like organisms (PPLO). There were put late on under the genus Mycoplasma by Nowak in 1929 and are now commonly called as mycoplasmas.

Principle

Mycoplasma in cell cultures or indeed elsewhere may be identified by both cultural and non-cultural methods. *Mycoplasma spp.* requires serum or ascetic fluid for growth, the essential requirements being protein, sterol and phospholipid. The isolation of mycoplasms is frequently done by the media containing antibiotics and other bacterial inhibitors.

Requirements

- Platinum loop
- Fresh clinical material
- Sterile Petri plates
- Constituents of PPLO agar medium

Arginine 0.1%

Glucose 0.1%

Phenol red 0.002%

Deoxyribonucleic acid 0.02%

Yeast extract 5%

Horse serum 10%

Mycoplasma broth base (BBL No. 11458).

Procedure

- Make PPLO agar media and pour into the sterile agar plate.
- Allow the plates to dry.
- Streak the plates with the clinical material obtained directly from a lesion with a sterile platinum loop.

- Incubate the plates aerobically as well as anaerobically for 48 hours at 36°C.
- Observe the results under low power objective and transmitted light for PPLO colonies.
- The PPLO appears colorless colonies 250-750 μm in diameter with a dense center.
- Colonies are further identified by staining and examining in situ.

Precautions

- Incubation atmosphere of 5-10% CO_2 in nitrogen is used as an alternative to conventional anaerobic incubation in hydrogen.
- Sterile distilled water should be used.
- Culture on solid media should be incubated.
- Aseptic conditions should be maintained.

11. EXAMINATION OF MICROBES IN FOODS

Introduction

There is food especially susceptible to microbial decomposition due to being rich in nutrients that support the growth of microbes. The contaminated food is often inedible and may also be the source of human disease, if the contaminating organisms are pathogens or toxin producers. Some of the important about food poisonings caused by bacteria such as *Botulism* (just because of *Clostridium botulinum*) *Clostridialers (clostridium perfringers), Enteric (Escherichia coli and Salmonella typhimurium) and Staphylococcal* (*Staphylococcus aureus*) are serious and some time is fatal. The sanitary food is a major concern with the examination of foods for the presence of pathogens.

Principle

There are various methods that might be used for the microbiological examination of food samples. The quantitative estimation of viable bacterial cells is present in a food sample by the serial dilution-agar plate technique and to determine the presence of coliform bacteria.

Requirements

- Food samples
- Inoculating loop
- Quebec colony counter
- Balance meter
- Bunsen burner
- Spatula
- Forceps
- Scalpel
- Sterile weighing paper squares contained in a Petri dish.
- Sterile Petri plates
- Sterile 1 ml pipettes
- Sterile water blanks (99 ml)
- Nutrient gelatin tube
- Litmus milk tube

- Lactose broth tubes
- Eosin methylene blue agar plate
- Brain-heart infusion agar deep tubes

Procedure

- Weigh 1 gm of food sample and transfer to the 9 ml water blank which contains glass beads.
- Shake well.
- Using a sterile 1- ml pipette, aseptically transfer 1 ml of the 1:10 sample into 99 ml water blank.
- Label the suspension ($1:10^3$) and shake the suspension.
- Transfer 1 ml suspension into another 99-ml water blank and label it $1:10^5$.
- Shake the suspension.
- Label the bottoms of the four sterile plates with the dilutions ($1:10^3$; $1:10^4$; $1:10^5$; $1:10^6$).
- Transfer 1 ml and 0.1 ml of the suspensions from $1:10^3$ suspensions into the plates marked ($1:10^3$; $1:10^4$) respectively.
- Repeat this process with $1:10^5$ dilutions and inoculate the plates labeled $1:10^5$ and $1:10^6$.
- Make five plates for each dilution.
- Prepare the brain heart infusion agar tubes and allow these to cool and pour melted and cooled agar to the plates containing the different dilutions.
- Rotate the plate gently to obtain a uniform distribution of microorganisms.
- Allow the plates to solidify.
- Keep the plates in the incubator at different temperatures for 48 hrs (10 °C, 25 °C, 35 °C, 45 °C and 55 °C).
- Inoculate the tube each of lactose broth, litmus milk and gelatin with a loopful of the material from the $1:10^6$ dilutions.
- Incubate these tubes at 35°C for 2 to 4 days.
- Aseptically prepare a four-way streak on eosin methylene blue agar plate from the $1:10^6$ dilutions.
- Incubate the plate at 37 °C for 48 hours.

- Arrange the plates incubated at each temperature from the lowest to highest dilution.
- Select the plates for the colony.
- Count the number of colonies on the plates selected.
- Determine the number of organisms per gram of the food sample by multiplying the number of colonies counted by the dilution factor.
- Record the number of colonies per plate and the number of organisms per gm of the food sample.
- Examine the eosin methylene blue agar plate and check the color of colonies.

Precautions

- Sterile distilled water should be used.
- Aseptic conditions should be maintained.
- Waste must be disposed in the waste container after completion of this test.

12. ALCOHOLIC FERMENTATION

Introduction

The term fermentation from the Latin word for boil was originally applied to the production of alcoholic beverages because the generation of carbon dioxide gas gives the fermenting liquid a brothy appearance. Most alcoholic beverages are manufactured by the metabolic activity of *Saccharomyces cerevisiae* or of closely related yeasts. It is a term used for any process in which chemical changes are brought about by the action of enzymes produced by living microbes in an organic substrate, whether carbohydrate, fat, protein or other types of organic material.

Principle

This general process is illustrated for fermentation: alcoholic fermentation and homolactic acid fermentation brought about yeasts and certain lactic acid bacteria respectively. Both of these fermentative processes utilize the Embden Meyerhof Paranas (EMP) pathway. In this case of alcoholic fermentation during wine production yeast uses sucrose to produce ethyl alcohol and carbon dioxide under anaerobic conditions. During the metabolism of glucose, pyruvic acid is first decarboxylated to form acetaldehyde, the reoxidation of NADH occurs concomitantly with the reduction of acetaldehyde to form alcohol. The yeasts ferment sugars in fruit juices and grain mash, producing wine, beer and other alcoholic beverages. The biochemical conversion of grape juice occurs when it is inoculated by yeast and incubated at 21-32°C for 3-5 days under aerobic conditions and followed by an anaerobic incubation period, the yeast cells enzymatically degrade the glucose and fructose in grape juice first to acetaldehyde and then to alcohol. The fermenting juice can have examined at 1-7 days' intervals for total acidity, volatile acidity and alcohol. The alcohol produced during the fermentation can be recovered using a distillation apparatus.

Requirements

- Bunsen burner
- Spatula
- Ring stand
- Glass beads
- Wire guaze

- Sodium hydroxide (0.1N)
- Grape juice broth culture of *Saccharomyces cerevisiae*
- Pasteurized commercial grape juice (500 ml)
- Phenolphthalein solution (1%)
- Sucrose (20 g)
- One-liter sterile Erlenmeyer flask
- Rubber stopper
- Sterile pipettes
- Burette
- Spatula
- Glaasine paper
- Alcoholometer
- Distillation apparatus

Procedure

- Take 500 ml of grape juice into 1-liter sterile flasks.
- Add 20 ml of sucrose and 50 ml of the culture of the yeast (10% starter culture) to the flasks.
- Close the flasks with a rubber stopper.
- Incubate the flasks for 21 days at 25 °C and determine total acidity, volatile acidity and alcohol at different time intervals.
- Add 20 gm of sucrose to the fermenting juice after 2 and 4 days of incubation.
- Determine the total acidity and volatile acidity by titration method at 1 to 7-day intervals.
  a) Take 10 ml of the fermenting juice to a flask.
  b) Add 10 ml of distilled water and 5 drops of phenolphthalein indicator (1%).
  c) Mix the contents.
  d) Titrate it against sodium hydroxide taken in a burette to the first persistent pink color.
  e) Note the volume used of sodium hydroxide.
  f) Repeat steps "a" to "e" using uninoculated grape juice.
- Determine the percentage of alcohol by dipping the alcoholometer in the fermenting juices at 7-day intervals.

- Assemble the distillation apparatus.
- Pour 100 ml of fermented grape juice into the distilling flask and add a few glass beads and put into the thermometer in the hole.
- Start the circulation of water through the condenser.
- Light the Bunsen burner and allow the liquid to drip in the receiver flask.
- Adjust the heat in such a way that the drops come at the rate of one per second.

Record the appearance, presence of turbidity or sediment and odor of the contents of the flask.

Total acidity and volatile acidity can be calculated by applying the formula:

$$\text{Total acidity (expressed as \% tartic acid)} = \frac{\text{Volume of alkali used (ml) x Normality of alkali x 75}}{\text{Weight of sample in gm (1 ml = 1 gm)}}$$

$$\text{Volatile acidity (expressed as \% acetic acid)} = \frac{\text{Volume of alkali x Normality of alkali X 6.0}}{\text{Weight of sample in gm}}$$

Record the total acidity and volatile acidity and % alcohol of the fermenting juice at 7-day intervals. The amount of alcohol is determined per 100 ml of the grape juice produced under laboratory conditions at 12 hr, 24 hr or 7-day intervals.

Precautions

- The thermometer bulb should always be placed just below the side-arm of the distilling flask.
- Joints must be airtight.
- We should check before lighting the Bunsen burner.

13. BREAD MAKING WITH THE USE OF YEASTS

Introduction

The fermentation process is an ancient time, so the production of bread is a very old process to discuss here in this exercise. There are commonly made by using *Saccharomyces cerevisiae* to dough from flour, water and some slats. The yeast aerobically respires in bread dough, produces amylases that act on starch and releases maltose that is later fermented and converted into carbon dioxide and water, rather than alcohol.

Principle

The rising of bread (Leavening) is achieved primarily through the release of carbon dioxide which makes it porous and spongy. The contribution to bread texture comes from kneading that breaks down flour proteins (gluten) and gives the dough elasticity. Baking the bread expands the trapped gas, causes further rising, solidifies the structure of the loaf, kills the yeast and inactivates the enzymes. In addition to the leavening of bread, the yeast produces volatile organic acids and alcohols that impart delicate flavors and aroma to bread.

Requirements

- Cylinders
- Dehydrated yeasts
- Flour
- Sugar
- Glass rods
- Mixing dishes
- Glass delivery tubes
- Bottles (250 ml)
- Two-hole rubber stoppers
- Bromothymol blue indicator solution (with one drop per liter of ammonium hydroxide).

Procedure

- Mix flour and sugar with water in a dish.

- Transfer the mixture into three reagents bottles.
- Set the apparatus and see the role of yeasts in bread making.
- Spread the yeast powder over the dough.
- Allow the fermentation of dough to take place (It can be tested by the change in color of the bromothymol blue indicator and these indicators are blue when alkaline and turn yellow (acidic) by the presence of carbon dioxide.
- To other reagents bottles, sprinkle yeast over the dough and mix these well with a glass rod.
- Incubate it for 30 minutes at room temperature.
- Puncture the dough with a glass rod at 5-minute intervals.
- Run a control where no yeast has been spread over the dough taken in a reagent bottle.
- Determine the time taken by the yeast to produce carbon dioxide in the reagent bottle.
- The yeasts had been spread over the dough without mixing and yeast had been mixed with dough and regularly punctured with the glass rod and compare the reactions with control.
- The fermentation of sugars present in the dough take place, carbon dioxide is evolved that turns the bromothymol blue indicator from yellow by changing it from alkaline to acidic.
- There will be more evolution of carbon dioxide where yeast is mixed well with the dough and the latter is regularly punctured.

Precautions

- Sterile distilled water should be used.
- Aseptic conditions should be maintained.

14. WINE PRODUCTION FROM GRAPES

Introduction

Wine is an alcoholic beverage arising from the fermentation of grape juice with many more fruits used for making wine such as apple, peaches, pears and plums. The production of wine is in the industry with the use of yeasts (*Saccharomyces cerevisiae and S. ellipsoideus*). This method is very common in U.S and Europe. The grape juice contains carbohydrates (glucose, fructose and pentose) and waters, nitrogenous compounds such as proteins and protein split products, acids, minerals, pigments and enzymes. Its concentration depends on grapes that determines the characteristics taste and texture of different wines. Grapes wines are two types red and white. Red wines are coming from the red or purple varieties and crushed grapes are fermented with their skin to allow the extraction of their color into the juice. White wines are prepared either from white or red grapes.

Principle

The juice containing 20-30% sugar will produce wines with an alcohol content of 10-15%. During the wine production, the yeasts convert glucose and fructose present in the juice enzymatically first to acetaldehyde and then to ethyl alcohol. The purpose of this exercise is to acquaint the students with the production of white wines from the juice of white grapes by a modified method. After one week, students will examine the wine produced in the laboratory for taste, odor, clarity, total acidity, volatile acidity and percent alcohol.

Requirements

- Glassine paper
- White grape juice
- Broth culture of yeasts at 25 °C (50 ml)
- Spatula
- Burette
- Ebulliometer
- Cylinder (10 ml)

- Rubber stopper fitted with a 2-inch glass tube plugged with cotton
- Flask
- Sodium hydroxide
- Phenolphthalein solution (1%)
- Sucrose (60 gm)
- Pasteurized white grapes juice (500 ml)

Procedure

- Pour 500 ml of pasteurized white grapes juice into a 1 liter sterilized flask.
- Add 20 gm of sucrose into the must.
- Now add 50 ml of the *S. ellipsoideus* grape juice broth culture into the flask.
- Close the flask with a sterilized stopper containing a cotton plugged air vent for producing aerobic conditions to the yeast.
- Incubate the inoculated fermenting wine flask at 25 °C for 21 days.
- Add 20 gm of sucrose to the fermenting wine after two days and then after four days of inoculation.
- Perform the following tests on the uninoculated white grape juice and samples taken from the fermenting wine at seven-day intervals for the taste, aroma and clarity.
- To check the total acidity expressed as % tartaric acid and volatile acidity expressed as acetic acid by the titration test for both the uninoculated juice and fermenting wine as described below:

To check total acidity

a) Take 10 ml of the grape juice in a 100 ml conical flask.
b) Add 10 ml of distilled water and 5 drops of phenolphthalein solution (1%).
c) Mix the contents well.
d) Titrate to the first persistent pink color with 0.1 N NaOH.
e) Calculate total acidity by the formula

$$\% \text{ tartaric acid} = \frac{\text{ml alkali} \times \text{normality of alkali} \times 7.5}{\text{Weight of sample (gm)}}$$

$$\text{Volatile acidity} = \frac{\text{alkali ml} \times \text{normality of alkali} \times 6.0}{\text{Weight of sample in gm}}$$

Precautions

- Healthy grapes should be collected.
- Grapes should be harvested when their sugar contents are between 15% and 25%.
- Always used sterile flasks
- Sterile distilled water should be used.

15. ANTIBIOTIC PRODUCING MICROBES

Introduction

The chemical compounds acting against life is called an antibiotic and it can be obtained either from natural sources or microbes or by a synthetic method. There are many different antibiotics are produced by bacteria, fungi and actinomycetes are therapeutically important. The simplest screening techniques for antibiotic producing microbes are the crowded plate procedure. There are many colonies producing antibiotic activity are indicated by an area of agar around the colony that is free of growth of other microbial colonies. Such a colony is sub-cultured to a similar medium and purified by streaking before making stock cultures.

Principle

Spore forming bacteria found in soil. They are characterized by the tough, leathery, frequently pigmented colonies and by their filamentous growth. They are capable of using complex organic materials such as carbon and energy sources and are involved in the breakdown of these products in the soil. This degradative ability makes these bacteria pivotal in the production of fertile soil for agriculture. They also give soil its characteristic smell by the production of a class of volatile low molecular weight compounds called geosmins.

Requirements

- Soil sample
- Nutrient agar

Beef extracts 3.0 gm

Peptone 5.0 gm

Agar Agar 15 gm

- Dissolve in 1000 ml of sterile distilled water and adjust the pH to 7.0

Procedure

- Take a soil sample and serially diluted from 10^{-1} to 10^{-7}.
- Make agar medium and autoclave before being used.

- Pour the agar medium into the agar plate and solidify all the plates.
- Plate 0.1 ml from each of the dilutions blanks onto the appropriate agar plates and spread the inoculums using a sterile L-rod-shaped spreader.
- Incubate the plates at 30°C for 5 to 7 days.
- Examine the plates and look for typical colonies producing antibiotics.
- The colonies are producing antibiotic activity that is indicated by a clear zone around the other microbial growth.

Precautions

- Sterile media should be used.
- Aseptic conditions should be maintained

16. PRODUCTION OF ETHANOL

Introduction

It is a complex biological process for the production of alcohol from sugar. It is widely used for many purposes. Glucose becomes ethanol, with the help of yeast in anaerobic conditions. It is a non-sterile process. The presence of undesirable microorganisms in the process cannot be completely avoided. The raw material is used for the fermentation process itself shows the presence of contaminating microorganisms.

Principle

The fermentation process of organic matter can produce liquor with 10-15% ethanol. The liquor is distilled to remove the water and other constituents and obtain high-grade ethanol. The exchanges associated with alcohol fermentation and distillation vary from plant to plant depending on technology, equipment used as well as operations practices. Ethanol is also called alcohol and the formula is C_2H_5OH.

Requirements

- Sulfuric acid
- Corn (meal, shelled and flour)
- Malt and yeast
- Sugar
- Conical flasks
- Measuring cylinder
- Beaker

Procedure

- Take 50 gm corn with 300 ml of preheated distilled water in a beaker.
- Add 1 to 2 drops of sulfuric acid to adjust pH to 5.
- Bring to a boil while mixing constantly.
- Boil for 15 minutes.
- Set aside and allow cooling while mixing constantly for 5 minutes.
- Mix 12 gm of ground malt with 100 ml of distilled water into the

flask.

- Keep aside for later use.
- Prepare the yeast and in the test tubes, mix 5.0 gm of yeast with 20 ml of warm 30°C water.
- Add a pinch of sugar and watch for bubbling to show yeast is active.
- Keep aside for later use.
- Carefully cool mash to 63°C while mixing at step 5.
- Add malt solution (step 6)
- Allow to cool and then mix in yeast solution (step 8).
- Cover the beaker with a glass or plastic top, or transfer the mixture into the flask.
- Keep out at room temperature for 10-14 days.

Preparation of malt

At least 14 days are required to prepare the malt.

Requirements

- Corn seed

Procedure

- Take 20 gm of soak corn seeds in a container of warm water for 24 hrs.
- Discard the water, and again add the warmer water and allow soaking for 15 minutes and drain water.
- Spread the soaked seeds on a wet towel and cover with another wet towel and then cover with plastic.
- Keep the towel moist for the seeds to sprout.
- When the sprouts reach a one-inch length, remove out sprout and seed and let it dry.
- Grind the sprout and seed for use in the mash called as malt.

Precautions

- Sterile distilled water should be used.
- Aseptic conditions should be maintained.
- It may be necessary to place the beaker in warm, cool and then ice water.

REFERENCES

1. **Abbott L., and Murphy D.,** 2003. Soil Biology fertility: A key to sustainable land use in agriculture. Kluwer Academic Publishers., 187-203.
2. **Adonizio A.L., Downum K., Bennett B.C., Mathee K.** 2006. Anti-quorum sensing activity of medicinal plants in southern Florida. J. Ethnopharm., **105**(3): 427-435.
3. Alexander, M. 1985. Introduction to Soil Microbiology (2ndE d.) Wiley Eastern.
4. **Bais H.P., Fall R., Vivanco J.M.,** 2004. Biocontrol of *Bacillus subtilis* against infection of *Arabidopsis* roots by *Pseudomonas syringae* is facilitated by biofilm formation and surfactin production. Plant Physiol., **134**: 307-319.
5. **Bais H.P., Tiffony L., Weir L.T., Perry L.G., Gilroy S., Vivanco J.M.,** 2006. The role of root exudates in rhizosphere interactions with plant and other organisms. Annu. Rev. Plant Biol., **57**: 233-266.
6. **Barber S.A.,** 1984. Soil Nutrient Bioavailability. John Wiley & Sons, NewYork.
7. **Barber S.A., and Lee R.B.,** 1974. The effect of microorganisms on the absorption of manganese by plants., New Phytol., **73**: 97-106.

8. **Belimov A.A., Hontzeas N., Safronova V.I., Demchinskaya S.V., Piluzza G., Bullitta S., Glick B.R.,** 2005. Cadmium-tolerant plant growth promoting rhizobacteria associated with the roots of Indian mustard (*Brassica juncea* L. Czern.). Soil Biol. Biochem., **37**: 241-250.
9. **Biró B., Köves-Péchy K., Tsimilli-Michael M., Strasser R.J.,** 2006. Role of beneficial microsymbionts on the plant performance and plant fitness. In: Mukerji KG, Manoharachary C, Singh J (eds) Microbial activity in the rhizosphere, Soil Biol. vol. 7. Springer-Verlag, Berlin., 265-296.
10. **Blake R.C., Choate D.M., Bardhan S., Revis N., Barton L.L., Zocco T.G.,** 1993. Chemical transformation of toxic metals by a *Pseudomonas* strain from a toxic waste site. Environ. Toxic and Chem., **12**: 1365-1376.
11. **Blaylock M.J., Salt D.E., Dushenkov S., Zakharova O., Gussman C.,** 1997. Enhanced accumulation of Pb in Indian mustard by soil-applied chelating agents. Environ. Sci. Technol., **31**: 860-865.
12. **Bottomley P.J., and Dughri M.H.,** 1989. Population size and distribution of *Rhizobium leguminosarum* biovar trifolii in relation to total soil bacteria and soil depth. Appl. Environ. Microbiol., **55**: 959-964.
13. **Bottomley P.J., and Maggard S.P.,** 1990. Determination of viability within serotypes of a soil population of *Rhizobium leguminosarum* biovar *trifolii*. Appl. Environ. Microbiol., **56**: 533-540.
14. **Brooks R.R.,** 1998. Geobotany and hyperaccumulators. In: Brook RR (ed) Plants that hyperaccumulate heavy metals. CAB International, Wallingford, pp 55-94.
15. **Cakmakci R., Donmez F., Aydın A., Sahin F.,** 2006. Growth promotion of plants by plant growth-promoting rhizobacteria under greenhouse and two different field soil conditions. Soil Biol. Biochem., **38**: 482-1487.
16. **Chris H.,** 2006." What is an Anaerobic biodigester?" http://manure.unl.edu/adobe/v7n10_01.pdf.
17. **Crowley D.E., Wang Y.C., Reid C.P.P., Szansiszlo P.J.,** 1991. Mechanism of iron acquisition from siderophores by microorganisms and plants. Plant and Soil., **130**: 179-198.

18. **Dobbelaere S., Vanderleyden J., Okon Y.,** 2003. Plant growth promoting effects of diazotrophs in the rhizosphere. Crit. Rev. Plant Sci., **22**: 107-149.

19. **Ehrlich H.L., 1990**. Geomicrobiology, 2nd edn. Dekker, New York, pp 646.

20. **Faraldo-Gomez J.D., and Sansom M.S.P.,** 2003. Acquisition of siderophores in gram-negative bacteria. Nat. Rev. Mol. Cell Biol., **4**: 105-116.

21. **Glick B.R.,** 1995. The enhancement of plant growth by free-living bacteria. Can J. Microbiol., **41**: 109-117.

22. **Glick B.R.,** 2001. Phytoremediation: synergistic use of plants and bacteria to cleanup the environment. Biotechnol. Adv., **21(3)**: 83-93.

23. **Glick B.R., and Penrose D.M., Li J.A.,** 2002. Model for the lowering of plant ethylene concentrations by plant growth promoting bacteria. J. Theor. Biol., **190**: 63-68.

24. **Gunatilaka A.A.L.,** 2006. Natural products for plant associated microorganisms: distribution, structural diversity, bioactivity and implication of their occurrence. J. Nat. Prod., **69**: 509-526.

25. **Gupta D.K., Rai U.N., Sinha S., Tripathi R.D., Nautiyal B.D., Rai P., Inouhe M.,** 2004. Role of *Rhizobium* (CA-1) inoculation in increasing growth and metal accumulation in *Cicer arietinum* L. growing under fly-ash stress condition. Bull. Environ. Contam. Toxicol., **73**: 424-431.

26. **Huang J.W., and Cunningham S.D.,** 1996. Lead phytoextraction: species variation in lead uptake and translocation. New Phytol., **134**: 75-84.

27. **http:**//www.tcd.ie/Biochemistry/IUBMB-Nicholson/swf/glycolysis.swf.

28. **http:**//www.northland.cc.mn.us/biology/Biology1111/animations/glycolysis.html.

29. **http:**//www.biocarta.com/pathfiles/h_glycolysisPathway.asp.

30. **http:**//www.accessexcellence.org/RC/VL/GG/out_Glycol.html.

31. **https:**//www.ibuf.coartuja.csic.es.

32. **https:**//www.environmentaleverage.com.

33. **https:**//www.astrosurf.com.

34. **https:**//www.britannica.com/science/bacteria/Bacterial-metabolism.

35. **https:**//www.lbl.gov.

36. **https:**//www.library.thinkquest.org.
37. **https:**//www.ecosys.uni-erlangen.de.
38. **https:**//www.miljolare.no.
39. **https:**//uh.edu/dtu/05-Transport-1-07.htm
40. **https:**//www.wasser-wissen.de.
41. **Kamnev A.A.,** 2008 Metals in soil verses plant-microbe interactions: Biotic and chemical interferences. In: Barka EA, Clément C. (eds) Plant-Microbe Interactions., 291-318.
42. **Kamnev A.A., Kovács K., Shchelochkov A.G., Kulikov L.A., Perfiliev Yu.D., Kuzmann E., Vértes A.,** 2006. Metal ions in biology and medicine, vol. 9, Alpoim MC, Morais PV, Santos MA, Cristóvão AJ, Centeno JA, Collery Ph (eds.) John Libbey Eurotext, Paris, p 220.
43. **Karthikeyan R., and Kulakow P.A.,** 2003. Soil plant microbe interactions in phytoremediation. Adv. Biochem. Eng. Biotechnol., **78**: 51–74.
44. **Keating M.H., Mahaffey K.R., Schoney R., Rice G.E., Bullock O.R., Ambrose R.B., Swartout J., Nichols J.W.,** 1997. Mercury Study Report to congress. EPA-452/R-97-003, Washington DC, Vol I Sec 3 pp 6-7.
45. **Khan A.G.,** 2005. Role of soil microbes in the rhizospheres of plant grown in trace metals contaminated soils in phytoremediation. J. Trace Elem. Med. Biol., **18**: 355-364
46. **Khan A.G.,** 2005a. Mycorrhizas and phytoremediation. In: Willey N (ed) Method in biotechnology-phytoremediation: methods and reviews. Humana Press, Totowa.
47. **Khan A.G.,** 2005b. Role of soil microbes in the rhizospheres of plants growing on trace metal contaminated soils in phytoremediation. J. Trace Elem. Med. Biol., **18**:355-364.
48. **Khan M.S., Zaidi A., Zaidi P.A., Mohammad O.,** 2009. Role of plant growth promoting rhizobacteria in the remediation of metal contaminated soils. Environ. Chem. Lett., **7**:1-19.
49. **Kingsley M.T., Metting F.B. Jr., Fredrickson J.K., Seidler R.J.,** 1993. In situ stimulation vs. bioaugmentation: Can plant inoculation enhance biodegradation of organic compounds. Paper 93-WA-89.04, Proceedings of the 86th Annual Meeting and Exhibition, Air and Waste Management Association, Pittsburgh, PA.

50. **Kirchhof G., Schloter M., Asmus B., Hartmann A.,** 1997. Molecular microbial ecology approaches applied to diazotrophs associated with non-legumes. Soil Biol. Biochem. **29**: 853-860.

51. **Kovacs K., Kamnev A.A., Mink J., Nemeth C.S., Kuzmann E., Megyes T., Grosz T., Medzihradszky-Schweiger H., Vertes A.,** 2006. Mössbauer, vibrational spectroscopic and solution X-ray diffraction studies of the structure of iron (III) complexes formed with indole-3-alkanoic acids in acidic aqueous solutions. Struct. Chem., **17**: 105-120.

52. **Kumar R., Bhatia R., Kukreja K., Behl R.K., Dudeja S.S., Narula N.,** 2007. Establishment of *Azotobacter* on plant roots: chemotactic response, development and analysis of root exudates of cotton (*G. hirusitum* L) and wheat (*T. aestivum* L). J. Basic Microbiol., **47**: 436-439.

53. **Langley S., and Beveridge T.J.,** 1999. Effect of O-side-chain-lipopolysaccharide chemistry on metal binding. Appl. Environ. Microbiol., **65:** 489-498.

54. **Lasat M.M.,** 2000. Phytoextraction of metals from contaminated soil: a review of plant-soil-metal interaction and assessment of pertinent agronomic issues. J. Hazard. Sub. Res., **2**: 5-25.

55. **Lima A.I.G., Pereira S.A.I., Figueira E.M.A.P., Caldeira G.C.N., Caldeira H.D.Q.M.,** 2006. Cadmium detoxification in roots of *Pisum sativum* seedlings: relationship between toxicity levels, thiol pool alterations and growth. Env. Exp. Bot., **55**: 149-162.

56. **Lloyd J.R.,** 2003. Microbial reduction of metals and radionuclides. FEMS Microb. Rev., **27**: 411-425.

57. **Lovley D.R., and Coates J.D.,** 2000. Novel forms of an aerobic respiration of environmental relevance. Curr. Opin. Microbiol., **3**: 252-256.

58. **Lucy M., Reed E., Glick B.,** 2004. Applications of free living plant growth promoting rhizobacteria. Ant van Leeuwen., **86**: 1-25.

59. **Lynch J.M.,** 1983. Soil biotechnology. Blackwell, Oxford.

60. **Lynch J.M.** 1990. Beneficial interactions between microorganisms and roots. Biotechnol Adv., **8**: 335-346.

61. **Mamaril J.C., Paner E.T., Alpante B.M.,** 1997. Biosorption and desorption studies of chromium (iii) by free and immobilized *Rhizobium* (BJVr 12) cell biomass. Biodegradation., **8**: 275-285.

62. **Mukerji K.G., Manoharachary C., Singh J., (eds.).** 2006. Soil Biology, vol. 7. Microbial activity in the rhizosphere., Springer, Berlin, Heidelberg.

63. **Narula N., Deubel A., Gans W., Behl R.K., Merbach W.,** 2006. Colonization and induction of Para nodules of wheat roots by phytohormone producing soil bacteria. Plant Soil and Environ., (Czech Repu) **52**(3): 119-129.

64. **Narula N., Kumar V., Saharan B.S., Bhatia R., Lakshminarayana K.,** 2005a. Impact of the use of biofertilizers on grain yield of wheat under varying soil fertility conditions and wheat-cotton rotation. Arch. Agron. and Soil Sci., (Germany) **51(1)**: 79-89.

65. **Narula N., Saharan B.S., Kumar V., Bhatia R., Bishnoi L.K., Lather B.P.S., Lakshminarayana K.,** 2005b. Impact of the use of biofertilizers on cotton (*Gossypium hirusetum*) crop under irrigated agro-ecosystem. Arch of Agron and Soil Sci., (Germany). **51**(2): 69-77.

66. **Narula N., Kothe K., Behl R.K.,** 2009. Role of root exudates in plant-microbe interactions Journal of Applied Botany and Food Quality., **82**: 122-130.

67. **Narula N., Bhatia R., Anand R.C., Gera R., Behl R.K.,** 2011. Interaction of *A. chroococcum* with wheat germ agglutinin (WGA). The Icfai Uni J Lif Sci (In Press).

68. **Narula N., Kothe E., Behl R.K., 2011**. Heavy metal resistance among *Azotobacter* spp: survival in garden and heavy metal contaminated soil using Indian Mustard the Icfai Uni J Life Sci (In press).

69. **Okon Y.,** 1985. *Azospirillum* as a potential inoculant for agriculture. Trends Biotechnol. 3: 223-228.

70. **Ontario 2006.** "Anaerobic Digestion Basics." 08/04. http://www.omafra.gov.on.ca/english/engineer/facts/04-097.htm.

71. **Osama O., Ibrahim,** Ph.D. Consultant Biotechnology Gurnee IL. 60031, bioinnovation04@yahoo.com.

72. **Pal M.K.I., Panwar B.S., Goyal S.I., Attila A., Laszlo M.,** 2010. Effect of lead (Pb) and mercury (Hg) on plant growth promoting rhizobacterias (PGPB) as *Azotobacter chroococcum* and *Pseudomonas spp.* using Indian Mustard (*Brassica juncea* L.) test crop. Proc Int Semi on Crop Sci for Sustain and Food Secur, Szeged, Hungary, June 1-3, 2010.

73. **Panwar B.S., Ahmed K.S., Mittal S.B.,** 2002. Phytoremediation of nickel contaminated soils by *Brassica* species. J. Environ. Dev Sustainability **1**:1-6.

74. **Park C.H., Keyhan M., Matin A.,** 1999. Purification and characterization of chromate reductase in *Pseudomonas putida*. Abs Gen Meet American Soc. Microbiol., **99**: 536-548.

75. **Peters N.K., Frost J.W., Long S.R.,** 1986. A plant flavone, luteolin, induces expression of *Rhizobium meliloti* nodulation genes. Sci., **233**: 977-980.

76. **Rao Subba K.S.,** 1974. ***Prospects of Bacterial Fertilization in India.*** Fertilser News, 19(12):32-38.

77. **RaoSubba, N.S.,** 2000. Soil Microbiology (Fourth Edition of Soil Microorganisms and Plant Growth). Oxford and IBH Publishing Co. Pvt. Ltd. New Delhi.

78. **Raskin I.,** 1996. Plant genetic engineering may help with environmenta cleanup. Proc Natl. Acad. Sci U SA **93**: 3164-3166.

79. **Renella G., Mench M., Landi L., Nannipieri P.,** 2005. Microbial activity and hydrolase synthesis in long-term Cd-contaminated soils. Soil Biology and Biochemistry. Soil Biol. Biochem., **37**: 133-137.

80. **Romheld V., and Marschner H.,** 1986. Mobilization of iron in the rhizosphere of different plant species. Adv. Plant Nutr., **2**: 155-204.

81. **Rosenblueth M., and Martinez-Romero E.,** 2006. Bacterial endophytes and their interactions with hosts. Mol. Plant Microbe Interact., **19:** 827–837.

82. **Ruggiero C.E., Boukhalfa H., Forsythe J.H., Lack J.G., Hersman L.E., Neu M.P.,** 2005. Actinide and metal toxicity to prospective bioremediation bacteria. Environ. Microbiol., **7(1)**: 88-97.

83. **Ruml T., and Kotrba P.,** 2003. Recent advances in marine biotechnology, vol. 8. In: Fingerman M, Nagabhushanam R (eds.) Bioremediation, Science Publishers Inc., Enfield (NH), USA, pp 81-155.

84. **Saitia S., and Narula N.,** 1989. Heavy metal resistance and hydrocarbon utilization in *Azotobacter chroococcum.* Ind. J. Microbiol., **29**(3): 213-215.

85. **Saitia S., Narula N., and Lakshminarayana K.,** 1989. Nature and role of plasmids in *Azotobacter chroococcum.* Biotechnol. Lett., (USA) **11**: 713-716.

86. **Salt D.E., Blaylock M., Kumar P.B.A.N., Dushenkov V., Ensley B.D., Chet I., Raskin I.,** 1995. Phytoremediation: A novel strategy for the removal of toxic metals from the environment using plants. Biotechnol., **13**: 468-475.

87. **Shakolnik M.Y.,** 1984. Trace elements in plants. Elsevier., New York. pp 140-171.

88. Shellee A. 2012. Morehead Introduction to Microbiology New England Institute of Technology.

89. **Shoebitz M., Ribaudo C.M., Pardo M.A., Cantore M.L., Ciampi L., Curá J.A.,** 2009. Plant growth promoting properties of a strain of *Enterobacter ludwigii* isolated from *Lolium perenne* rhizosphere. Soil Biol. Biochem., **41(9)**: 1768-1774.

90. **Sigel A., and Sigel H.,** 1998. Metal ions in biological systems, vol. 35. Iron transport and storage in microorganisms, plants, and animals. Marcel Dekker, New York.

91. **Sigel H., and Sigel A.,** 1996. Metal ions in biological systems, Vol. 32, Interaction of Metal Ions with Nucleotides Nucleic Acids, and their Constituents, Marcel Dekker, New York.

92. **Simon L., Biró B., Balázsy S.,** 2008. Impact of Pseudomonads and ethylene on the cadmium and nickel rhizofiltration of sunflower, squash and Indian mustard. Commu in Soil Sci and Plan Anal., **36**(15-16): 2440-2455

93. **Simon L., Széles É., Balázsy S., Biró B.,** 2007. Selenium phytoextraction, speciation and microbial groups in Se contaminated soils. Proceeding, ICOBTE, 9-2007.

94. **Somers E., Vanderleyden J., Srinivasan M.,** 2004. Rhizosphere bacterial signaling: A love parade beneath our feet. Crit. Rev. Microbiol., **30**: 205-240.

95. **Sugawara M., Okazaki S., Nukui N., Ezura H., Mitsui H., Minamisawa K.,** 2006. Rhizobitoxine modulates plant–microbe interactions by ethylene inhibition. Biotechnol. Adv., **24**: 382-388.

96. **Teplitski M., Chen H., Rajamani S., Gao M., Merighi M., Sayre R.T., Robinson J.B., Rolfe B.G., Bauer W.D.,** 2004. *Chlamydomonas reinhardtii* secretes compounds that mimic bacterial signals and interfere with quorum sensing regulation in bacteria. Plant Physiol., **134**: 137-146.

97. **Tugarova A.V., Kamnev A.A., Antonyuk L.P., Gardiner P.H.E.,** 2006. Metal ions in biology and medicine, vol. 9. John Libbey Eurotext, Paris, 242.

98. **Von Uexkull H.R., Mutert E.,** 1995. Global extent, development and economic impact of acid soils. Plant Soil., **171**: 1-15..

99. **Walker T.S., Bais H.P., Grotewold E., Vivanco J.M.,** 2003. Root exudation and rhizosphere biology. Plant Physiol., **132**: 44-51.

100. **Wang P.C., Mori T., Komori K., Sasatsu M., Toda K., Ohtake H.,** 1989. Isolation and characterization of an *Enterobacter cloacae* strain that reduces hexavalent chromium under anaerobic conditions. Appl. Environ. Microbiol., **55**: 1665-1669.

101. **Wani P.A., Khan M.S., Zaidi A.,** 2007a. Cadmium, chromium and copper in green gram plants. Agron. Sustain. Dev., **27**: 145-153.

102. **Wani P.A., Khan M.S., Zaidi A.,** 2007b. Impact of heavy metal toxicity on plant growth, symbiosis, seed yield and nitrogen and metal uptake in chickpea. Aust. J. Expt. Agric., **47**: 712-720.

103. **Wani P.A., Khan M.S., Zaidi A.,** 2007c Effect of metal tolerant plant growth promoting *Rhizobium* on the performance of pea grown in metal amended soil. Arch. Environ. Conatm. Toxicol. doi:10.1007/00244-9097-y.

104. **Wani P.A., Khan M.S., Zaidi A.,** 2008a. Effect of heavy metal toxicity on growth, symbiosis, seed yield and metal uptake in pea grown in metal amended soil. Bull Environ Contam Toxicol doi:10.1007/s00128-008-9383-z.

105. **Wani P.A., Khan M.S., Zaidi A**., 2008b. Chromium reducing and plant growth promoting *Mesorhizobium* improves chickpea growth in chromium amended soil. Biotechnol. Lett., **30**: 159-163.

106. **White P.J.,** 2003. Ion transport. In: Thomas B, Murphy DJ, Murray DJ (eds) Encyclopedia of plant sciences. Academic press, London, pp 625-634.

107. **Wu S.C., Cheung K.C., Luo Y.M., Wong M.H.,** 2006. Effects of an inoculum of plant growth promoting rhizobacters on metal uptake by *Brassica juncea*. Environ. Pollu., **140**: 124-135.

108. **Yang X., Feng Y., He Z., Stoffella P.J.,** 2005. Molecular mechanisms of heavy metal hyperaccumulation and phytoremediation. J Trace Elem. Med. Biol., **18(4)**: 339-53

109. Yasmin S., Bakar M.A.R., Malik K.A., Hafeez F.Y., 2004. Isolation, characterization and beneficial effects of rice associated PGPB from Zanzibar soils. J. Basic Microbiol., **44**:241-252.

110. Zachara J.M., Fredrickson J.K., Smith S.C., Gassman P.L., 2001. Solubilization of Fe(III) oxide-bound trace metals by a dissimilatory Fe(III) reducing bacterium. Geoch. Cosomo. Acta., **65**: 75-93

111. Zaidi A., and Khan M.S., 2006. Co-inoculation effects of phosphate solubilizing microorganisms and *Glomus fasciculatum* on green gram-*Bradyrhizobium* symbiosis. Turk J. Agric. for., **30**: 223-230

112. Zaidi S., Usmani S., Singh B.R., Musarrat J., 2008. Significance of *Bacillus subtilis* strains SJ-101 as a bioinoculant for concurrent plant growth promotion and nickel accumulation in *Brassica juncea*. Chemo., **64**:991-997

INDEX

D

E

F

G